P9-DNZ-046

A BRIGHT FUTURE

A BRIGHT FUTURE

How Some Countries Have Solved Climate Change and the Rest Can Follow

JOSHUA S. GOLDSTEIN
AND STAFFAN A. QVIST

PUBLICAFFAIRS
NEW YORK

Cover design by Pete Garceau
Cover image © Getty Images

PublicAffairs
Hachette Book Group
1290 Avenue of the Americas, New York, NY 10104
www.publicaffairsbooks.com
@Public_Affairs

Printed in the United States of America

First Edition: January 2019

Published by PublicAffairs, an imprint of Perseus Books, LLC, a subsidiary of Hachette Book Group, Inc. The PublicAffairs name and logo is a trademark of the Hachette Book Group.

Print book interior design by Jouve.

Library of Congress Control Number: 2018960023

ISBNs: 978-1-5417-2410-5 (hardcover), 978-1-5417-2409-9 (e-book)

LSC-H

10 9 8 7 6 5 4 3 2 1

Contents

Foreword

By Steven Pinker

Steven Pinker is Johnstone Family Professor of Psychology at Harvard University and the author of ten books, most recently Enlightenment Now: The Case for Reason, Science, Humanism, and Progress.

FEW BOOKS CAN credibly claim to offer a way to save the world, but this one does. Climate change is the most pressing issue facing humanity today, and the major responses to it—denial on the Right, abandoning industrial capitalism on the Left, and installing solar panels and wind turbines in the center—will not avert a possible catastrophe.

Joshua Goldstein and Staffan Qvist have written the first book on dealing with climate change that is grounded in reality. They have scrutinized the science and technology, done the math, surveyed the energy landscape around the world, and, most important, considered the political context—since a technically feasible solution is pointless if no one will adopt it.

A Bright Future was not written by polarizing politicians or icons of environmentalism, and therein lies its strength.

Surveys have shown that people resist accepting the reality of human-made climate change not because they are scientifically ignorant but because they associate the claim with the political Left and with communitarian and puritanical values regarding modern life with which they do not sympathize. And the traditional environmental movement has treated climate change like other environmental problems and advocated measures like conservation and small-scale energy generation that are not commensurate with the magnitude of the threat to human well-being.

Goldstein comes to the problem of climate change from his expertise in a different global-scale topic: international relations, including war and peace. The author of two prize-winning books on war and of the most popular textbook in his field, he is accustomed to thinking of existential threats and, crucially, how they have been tamed. (His discovery that war has declined in recent decades, presented in *Winning the War on War*, was a major inspiration for my own book *The Better Angels of Our Nature: Why Violence Has Declined.*) Qvist is a hardheaded expert on energy technologies around the world.

These two heavyweights have taken a fresh look at climate change, with no agenda other than to figure out how to solve the problem. They begin with some undeniable truths. Energy-hungry industrialization has been good for humanity, lifting people out of abject poverty and allowing them to live long, healthy, comfortable, and stimulating lives.

Today's poor have a right to enjoy that progress in their turn. But the world now faces a crisis as its growing energy use, almost entirely with cheap and convenient fossil fuels, threatens disastrous consequences for the Earth's climate.

Advances in policy and technology promise the possibility that we can have more energy with less pollution—but will these changes kick in fast enough to win the race against a possible catastrophe? The advances would have to allow us to replace the foundation of the world economy—the fossil fuels supplying 85 percent of its energy—with new energy sources that do not emit carbon. This would have to happen at a breakneck pace, most of it by the middle of this century.

Humanity has never faced a problem like this, and the popular solutions are not up to the challenge. Like many right-thinking people, I have done my part to encourage small personal sacrifices—for example, by cheerfully posing for student-sponsored posters encouraging the Harvard community to unplug chargers and take shorter showers. But anyone who looks at the numbers knows that such feel-good measures will leave only the faintest scratch.

A Bright Future is climate change for grown-ups. Rather than starting from baby steps and hoping these add up, it starts from where we need to end up and asks how we can get there. The obvious lodestar should be the few countries that *have* gotten there, or very close: the ones that have rapidly switched from fossil fuels to clean energy without taking a vow of poverty. We know that their approach can succeed because it already has. Since energy is good and carbon emissions are bad, the number we should track is grams of carbon emitted per kilowatt-hour (kWh) of electricity generated. By that measure, Sweden, France, and Ontario come in at one-tenth the world average, a level that would solve our problem if all the other countries matched their performance. And needless to say, these places are not squalid poorhouses but among the most pleasant places to live on Earth. How can we learn from their success? This is the practical question answered in *A Bright Future*.

Among the myriad responses inspired by the threat of catastrophic climate change, the one that I encounter most often these days is a helpless (and potentially self-fulfilling) fatalism: the planet is cooked, and there's nothing we can do about it but mourn for our future and enjoy life while we can. *A Bright Future* offers a constructive alternative. Human ingenuity got us into our predicament, and human ingenuity can get us out of it. By showing how we can solve the problem, *A Bright Future* is the most important book on climate change since *An Inconvenient Truth* and the perfect book for our time—one that could, quite literally, save the world.

DECARBONIZATION

The task before humanity is to
rapidly shift from CO_2-emitting
fossil fuels, which now provide
85 percent of the world's
growing energy needs,
to clean energy.

CHAPTER ONE

Climate Won't Wait

IF YOU THINK climate change is a serious problem, we have bad news: it's worse than you think.

We all see the graph of carbon pollution—emissions of carbon dioxide, CO_2—going up year by year and the graph of global temperature rising year by year. So it's natural to feel that if we just stopped the rise in CO_2 emissions, the temperature would also stop rising. Stopping the rise in emissions is within reach; it's what the Paris Agreement would do if the United States rejoined it and every country in the world achieved its goals under the treaty. But that would not stop global warming.[1]

Think about it: even if emissions stopped rising, we would still be putting CO_2 into the atmosphere at today's high rate, and the concentration of atmospheric CO_2 would keep going up. The CO_2 concentration has already risen from about 280 parts per million (ppm) before industrialization to about 410 ppm currently. Since CO_2 stays in the atmosphere for hundreds of years and nobody has yet invented a cheap and effective way to remove it, every ton we put into the air will stay there a long time.

At today's rate, every year the world puts about 35 billion tons of new CO_2 into an atmosphere already overloaded. That much CO_2 weighs about as much as 15 billion Ford Explorer SUVs, 2 for every human being on the planet, added every year. Other greenhouse gases, primarily unburned methane, contribute half again as much warming effect.[2] The Paris Agreement, if successful, would continue putting that much additional carbon into the atmosphere every year.[3] We need instead to quickly *reduce* that rate toward zero, but no plan currently in play does that effectively.

In fact, in the twenty-first century, the fastest-growing energy source in the world is coal, the most CO_2 intensive and toxic of the fossil fuels. Coal use has spiked faster than ever since 2001.[4] China alone in just five years, 2001–2006, *doubled* its already huge coal consumption. President Trump's promise in 2017 to end the US "war on coal" and ramp up the industry's growth is only the latest, minor, chapter in the story. The growth of coal is occurring mostly in poorer countries, because coal is cheap, while the fracking revolution in the United States has led to the steady replacement of coal by even cheaper methane (natural gas).

Together, the fossil fuels—coal, oil, and methane—supply 85 percent of the world's energy and are the main source of CO_2 emissions.[5] That percentage has to be quickly, in just a few decades, reduced to near zero, a herculean task on a global scale. This process of shifting off of fossil fuels is called *decarbonization.*

Even if we immediately stopped putting any new carbon in the atmosphere, today's CO_2 concentration of 410 ppm would cause temperatures to keep rising, although more slowly.[6] It would take a long time to bring temperatures back, but we'd have a good chance to head off the worst of the crisis. But until we stop adding more carbon, and not

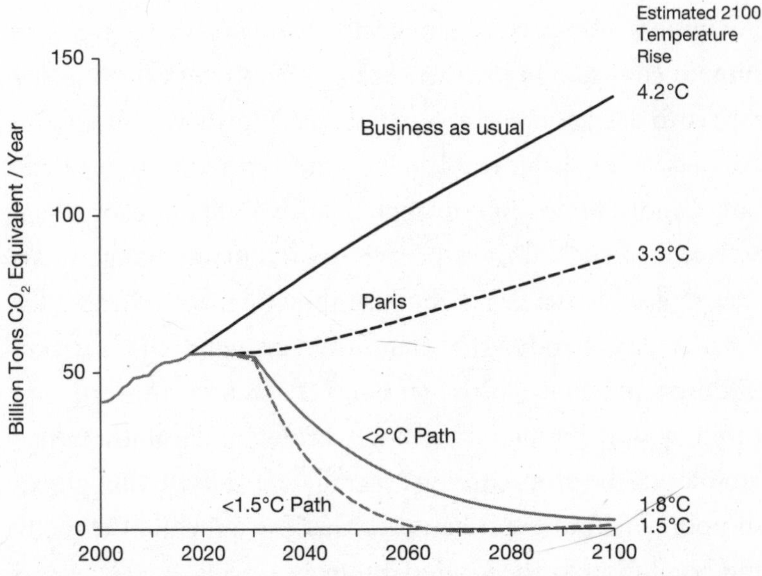

Figure 1. CO_2 emissions and warming. *Source*: Adapted by permission from Climate Interactive.

just flatten out the *rate* at which we keep adding carbon every year, we don't have a hope.

To accomplish a rapid decarbonization in the next couple of decades, the world will need to cut emissions in half for each of the next few decades, according to one road map.[7] What matters most right now is how fast the world can accomplish this. Carbon emissions today affect future climate outcomes, and the process is not linear. You might think that a delay of a decade or two in phasing down fossil emissions would merely advance bad climate outcomes by a decade or two, but it's much more serious than that.

Two Kinds of Change

To understand why, we need to distinguish two types of climate change. One is the kind of effect we already see or expect to see soon and that climate scientists warned about decades ago—rising sea levels, more frequent large hurricanes, more floods and droughts and wildfires, record heat waves, and such. When people say "climate change is already here," these are the effects they refer to.[8] Hurricanes Katrina and Sandy, the California drought, the Russian wildfires and those in the western US states, the European flooding, and the super typhoon in the Philippines are all recent examples of extreme weather of the type that global warming makes more likely. No single event can be directly tied to global warming, but the overall pattern fits what a warming planet produces.

But these events are merely inconvenient and expensive in the big picture. "Climate change is here now" does not convey the reality that climate change in the coming years will be far, far worse than today's extreme weather.

The second type of climate change is the potential "tipping points" that cause truly catastrophic shifts. These are still uncertain, and we may not know we have gone over the edge into irreversible shifts until it is way too late to do anything about it. We *might* have more time or turn out to be okay, but this is a terrible gamble to make. To invite a substantial possibility of catastrophic tipping points would be extremely irresponsible.

One catastrophic potential is for large sea-level rise much faster than is generally expected. We currently measure sea-level rise in inches, and we can adapt (at some expense) by building seawalls and moving infrastructure around. But some climate models show the potential for sea level to rise by 10 feet or more in this century. That's a game changer, in a world where most of the major cities are located on the coasts.

In New York, the average high tide, twice a day, would be higher than the Hurricane Sandy flooding. Downtown would be underwater. In Boston, Logan Airport would be submerged, as would Harvard and the Massachusetts Institute of Technology (MIT). New Orleans and Miami would be way underwater. Same for San Francisco's airport. (The group Climate Central has created photo illustrations of

Figure 2. Boston's Back Bay neighborhood under a scenario of 12 feet of sea-level rise. *Graphic*: Courtesy of Nickolay Lamm / Climate Central.

these locations.)[9] Outside the United States, the outcomes would be worse. Coastal cities across Asia with hundreds of millions of inhabitants could be severely impacted.[10] And in West Africa, tens of millions live in vulnerable coastal regions.

Most of the world's ice, some 7 million cubic miles of it, is in the ice sheet covering Antarctica in the South. But another ice sheet, on Greenland, contains more than a half million cubic miles' worth. To put these quantities in perspective, if all the Antarctic ice melted, sea levels would rise by 200 feet, and if the Greenland sheet all melted, the sea would rise by 20 feet.[11]

So far, the North of the Earth has been melting faster than the South, with the relatively thin sea ice in the Arctic Sea having shrunk by a third in the past twenty-five years. The summer of 2016 registered record high Arctic temperatures, 20°C above normal.[12] (The world as a whole is currently about 1°C above preindustrial levels.) This Arctic thaw is quite dangerous in its own right, because it can shift global weather patterns such as those driven by the jet streams. And multiple "positive feedback loops" are making the problem accelerate. Melting sea ice means less reflection of sunlight, which means warmer Arctic water and less ice. Melting permafrost on land is releasing methane gas, which increases global warming and melts more permafrost.

One possible disaster resulting from warmer temperatures in northern areas is a potential tipping point associated with the Greenland ice sheet. The "Atlantic conveyor belt" consists of warm water moving up the eastern coast of North America as the Gulf Stream and then sinking 10,000 feet near Greenland and moving back to the equator, where it warms and rises again. Large amounts of freshwater entering the North Atlantic off Greenland as the ice melts could shut down the conveyor belt because freshwater does not sink like saltwater. This could trigger an ice age in North America and Europe—an ironic consequence of global warming but one associated in the past with the conveyor-belt shutdown. Climate scientists worried about this possibility several decades ago and then decided about a decade ago that it was very unlikely, but they have now begun to worry about it again.[13]

The difference is dramatic between the inconvenience and expense of today's climate change and the catastrophe of climate tipping points in the upcoming decades or centuries. For example, Boston's winter of 2015 saw record snowfall, as freakish weather proliferated globally in a changing climate. With 6–8 feet of snow on the ground for weeks, streets became impassable, people couldn't get to work, and businesses shut down. The economic costs may have approached $1 billion.[14] That was inconvenient.

But imagine Boston under a mile-thick sheet of ice, as it was 12,000 years ago (a short time in geological scale). That's beyond inconvenient; it's "game over." New Orleans was devastated by Hurricane Katrina, which was tragic but temporary. But imagine New Orleans permanently under 10 feet of water. Imagine that California's five-year drought hadn't ended in 2017 but continued indefinitely, eventually depleting both reserves and aquifers, leaving an uninhabitable desert.

A *New York* magazine article in 2017 painted the picture of worst-case climate outcomes if we don't solve the problem and don't get lucky. The title, "The Uninhabitable Earth," sums it up. The author reminds us that several past "mass extinction" events in Earth history were caused by greenhouse gases that warmed the planet and that the worst of these killed 97 percent of life on Earth.[15]

By the way, many warnings about climate change highlight the potential for violent conflict, but that is not a main concern, certainly not compared to a new ice age or rapid sea-level rise. To be sure, a world of changing climate could amplify larger-scale migration and fights over natural resources.[16] These are real concerns and increasingly the focus of sustained policy attention.[17] But these would take place in a world where war and violence have broadly declined over several generations.[18] For instance, a well-publicized estimate that armed conflicts could increase by 50 percent as a result of climate change[19] would still mean levels of conflict well below those of the Cold War. Also, refugees are more often a consequence than a cause of major armed conflicts, and natural disasters can not only fuel conflict but also sometimes reduce it, as happened after the 2004 tsunami in Aceh, Indonesia, and the 2015 earthquake in Nepal.[20] Claims that climate-induced drought fueled the civil war in Syria are probably overstated.[21] Any increase in war, of course, is bad, but war is not the main outcome to worry about. The climate tipping points that would destabilize the planet's ecosystem must remain our main focus.

We don't know whether tipping points will come into play and trigger truly catastrophic outcomes, or when. One recent study gives current policies only a 5 percent chance of keeping global temperatures below 2ºC, the UN target for reducing the likelihood of catastrophic outcomes.[22] But many climate scientists feel that even the UN target of

2°C for global warming is not at all safe.[23] Any reasonable approach must recognize that catastrophes of this nature should be prevented with great vigor and dispatch even if their timing and eventuality are not certain.

A Slow-Motion Asteroid

Climate change is, therefore, not an environmental issue but an existential one. It is the slow-motion equivalent of a large asteroid heading toward Earth. Imagine that scientists discovered such an asteroid far off in space, headed our way. Their best guess was that it would probably hit us, but whether it destroyed just a few cities or all life on Earth was still uncertain. It might even miss altogether, although only 3 percent of scientists believed that.

What would we do? Clearly, and especially if the impact was just a few years away, we would mobilize the full capabilities of the world's countries, especially our military forces and budgets, to meet the threat. We would put the brightest minds on the task of figuring out a solution, and we would get out there to meet and deflect the asteroid at the earliest feasible date. With every day of delay, the asteroid would come closer and the task of deflecting it would become more challenging.

We would not argue about whether potential solutions were too technological and not "natural" enough. We would not complain that big corporations were going to make huge profits off the project (of course they would). We would not put our efforts into organizing for social justice on the grounds that the asteroid would probably affect poor people more than rich people (of course it would). We would not, most of us, go into denial or declare the end of Earth to be God's will. We would get out there and give the thing a good bump to save our planet.

But suppose, instead of a few years, the asteroid was not going to hit for a few decades—let's say, on its next orbit around the sun instead of this one. It would still be true that now is the cheapest, safest, most effective time to get out there and change its course. But we might lose the urgency and get distracted. By the time we tried to change the asteroid's trajectory, it might be too late to succeed.

Figure 3. Children and future generations will be most affected. Here, annual floods in Indonesia, 2013. *Photo*: Kate Lamb / VOA via Wikimedia Commons (CC BY-SA 3.0).

This is the trouble with climate action: Measures taken in the short term, especially in the next decade or two, will determine the long-term outcomes, but the pain and costs of the long-term outcomes will not be felt until decades later. It's act now, benefit later. Those most affected have no voice and no vote because they are very young or haven't yet been born.[24] In fact, a group of young Americans, who will bear the costs, have sued the federal government for the right to a stable future climate.[25]

Unfortunately, climate change has become a partisan issue in the United States. Conservatives deny any problem, and liberals too often fold the issue into a wider agenda of ending capitalism, globalization, inequality, and injustice. Author Naomi Klein calls climate change a "historic opportunity" to achieve these long-standing leftist goals.[26] As environmentalist George Marshall argues, climate change calls for a narrative of common purpose (humans need to rise to the challenge of climate change together), but people are more motivated by an "enemy narrative" (for example, the evil corporations are to blame). This leads many people to just ignore climate change even though they know it's a serious problem.[27]

The authors of this book—a political scientist and an energy engineer—share a deep concern for climate change and an alarm that the world is falling far short of what is needed to address it. We vigorously support today's popular solutions such as solar power,[28] wind power, and energy efficiency. But, as we will see in later chapters, these solutions simply do not add up fast enough to do what's needed. And if climate solutions have to wait for the end of capitalism, we're all in very deep trouble indeed.

Timing Is Everything

Just to meet the Paris Agreement targets, action before 2020 is critical. In 2017 a long list of climate policy leaders called for sweeping and almost immediate changes to begin reducing CO_2 emissions by 2020. "If we delay," they warned, "the conditions for human prosperity will be severely curtailed."[29]

Computer simulations developed at MIT[30] show the effect of timing in terms of when carbon emissions peak and how fast they decline. The model makes two things clear. First, almost regardless of anything we do, the world will pass 1.5°C around 2040. The Paris Agreement urged us to try to stay below that level if possible, but in truth that chance has passed. Second, what we do in the next ten years to peak and rapidly reduce emissions determines what happens in the second half of the century. A rapid decarbonization starting in 2020 means staying within the 2°C target that the United Nations has established as the upper limit. Anything less aggressive means blowing past that level in a few decades, as we will blow past 1.5°C around 2040. And business as usual means a 4.5°C increase by 2100.

If we peaked emissions immediately and let them continue at today's level, as the Paris Agreement would do, temperatures would still rise by more than 3°C by the end of the century. But if instead we reduced emissions by about 2–3 percent each year starting in 2020,[31] total emissions from the energy sector would go below zero by 2065 and global temperature rise would reach only 2°C by about 2070 and then stay there.[32] This kind of reduction in carbon pollution, about 30 percent per decade, would be the needed rapid decarbonization. Fifty percent per decade would be better, but 30 percent would work.

As this book will show, such a goal is achievable, but not the way we are going about things now. Nor does the idea of cobbling together a series of "climate stabilization wedges"—each a step in the right direction using existing technology—get us to the goal.[33] In the fifteen years since these wedges were proposed, little progress has been made on them individually, and overall progress has also not materialized. We need to look at the big picture and not just steps in the right direction.[34]

Focus on Electricity

This book focuses on electricity generation. Fossil-fuel carbon emissions come mostly from three main sectors of the economy—electricity generation, transportation, and heat (for buildings and industrial processes). Changes in land use, agriculture, and forests are also important for the climate, as is production of both steel and cement. Our primary focus is on phasing out fossil fuels used for electricity, because this is the quickest and most far-reaching way to reduce emissions. Emission reductions in transportation and heating will probably, to a large extent, involve electricity,[35] so clean electric power becomes all the more important in displacing fossil fuels.

By no means does this suggest that other aspects of greenhouse gas reduction can be ignored. We need to move from deforestation to reforestation, to change agricultural practices broadly, to implement energy-efficiency measures in all our vehicles and buildings, and so forth. We support all these efforts, but in this short book our primary focus is on rapidly decarbonizing electricity generation.

The units we use to measure electricity can sound technical, so here is a quick summary. The watt (W) is the basic unit of power—how much energy is produced or consumed in a unit of time. The old standard incandescent bulb was a 100W bulb. More often we will speak in kilowatts (kW, a thousand watts). A kW used for one hour makes a kilowatt-hour, which is the unit of energy you will find on your electric bill. In the United States, the average retail price of electricity is about 10 cents/kWh,[36] though it can be double that in places. The rate includes somewhat more than half for generation and the rest for transmission and distribution. A good wholesale price for electricity generation is around 5 cents/kWh, while anything around 10 cents or even 20 cents becomes uncompetitive economically. Such numbers come into play in later chapters. On a larger scale, the unit for measuring a typical power plant is a gigawatt (GW, a billion watts). Power production is measured in terawatt-hours (TWh, a billion kWh). Unless otherwise noted, since we are writing about electricity, units of generation capacity such as a GW refer to the electric production, often called a GWe, rather than the heat energy produced in generating that electricity.

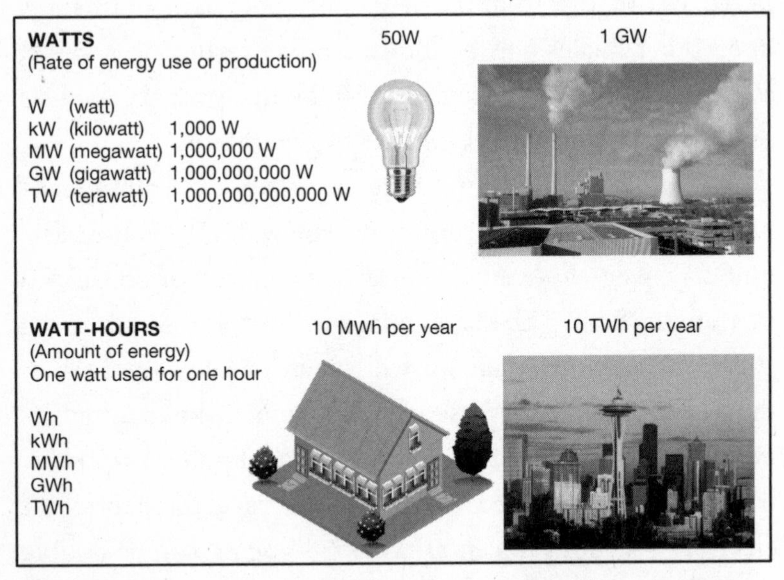

Figure 4. Units of electricity, with order-of-magnitude examples.
Source: Authors' graphic; Pickit photos.

So, to summarize, this is the situation: In order to avert future catastrophe, we need to rapidly reduce global carbon emissions by 2–3 percent a year, starting almost immediately.

The world as a whole has never done this before, but several individual countries have. They are the only models we have that prove the possibility of rapid decarbonization. We will examine these cases and then consider whether there are other ways to reach the same result.

CHAPTER TWO

What Sweden Did

ONE COUNTRY STANDS out for its success in rapid decarbonization. From 1970 to 1990, Sweden cut its total carbon emissions by *half* and its emission per person by more than 60 percent. At the same time, Sweden's economy expanded by 50 percent, and its electricity generation more than doubled.[1]

It all started in the late 1960s, and not because of concerns about climate change. By the end of the sixties, Sweden had started to halt the expansion of hydropower, in order to protect some of its last remaining undammed rivers.[2] At the time, it was not clear what energy source could be used to cover the ever-increasing demands for electricity, but oil was the most likely candidate. However, the oil crises

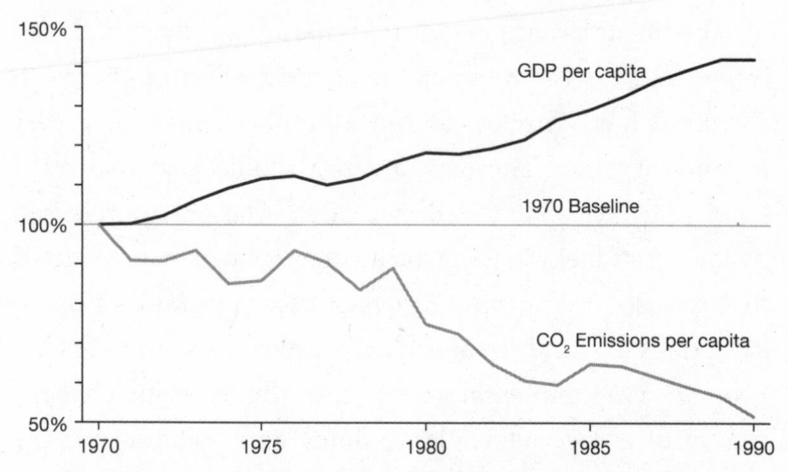

Figure 5. Sweden's gross domestic product and CO_2 emissions, 1970–1990. *Data source*: World Bank. GDP in constant dollars, not purchasing-power-parity (PPP) adjusted.

of 1973 and 1979, which caused huge price spikes and supply disruptions, convinced the Swedes to develop an alternative to imported fossil fuels.

Instead of expanding use of fossil fuels such as oil to cover its growing electricity demands, Sweden built a series of power plants using a new energy source called *kärnkraft*. This source is carbon-free like hydropower, cheaper than imported oil, much less harmful to health than coal, and incredibly concentrated. One pound of kärnkraft fuel produces the same energy as more than 2 million pounds of coal![3] The amount of toxic waste produced in the electricity generation process is also many thousands of times less than using coal and even much less than methane (natural gas).

The big difference between kärnkraft and fossil fuels (or renewables) is the tremendous concentration of energy in kärnkraft fuel. The fuel to run a kärnkraft unit for a year fits onto a truck. The fuel to run a similar-size coal plant for a year fills 25,000 railroad cars.[4] The energy released by kärnkraft fuel weighing as much as a single penny equals that released by burning 5 *tons* of coal.[5] A similar disparity applies to waste streams from kärnkraft and from fossils. From an environmental perspective, this fantastic concentration of energy allows less mining and pollution for the same amount of energy.[6]

Kärnkraft provided Sweden with a glut of reliable and cheap electricity. It not only enabled the country to expand its use of energy and electricity very rapidly without the increased use of fossil fuels (as was the case in most of the rest of the world), but also enabled Sweden to retire its existing fossil-fueled energy supply. The kärnkraft expansion, along with the development of biomass and waste-fueled district heating systems,[7] drove many previously fossil-fueled activities to switch to clean energy. The total energy supply from oil products dropped by more than 40 percent, and in the same time period the use of electricity for heating expanded fivefold.[8]

Sweden built a dozen kärnkraft units, grouped on just four sites, in the 1970s and '80s and even built a couple of units in neighboring Finland at the same time. At the peak of kärnkraft's rollout in the 1970s, Sweden had about one large unit under construction per million citizens. The same rate applied in China or India today would mean a concurrent construction of more than a thousand units in each country (as we will discuss later). Eight of the plants built in Sweden continue to operate today (as do the two in Finland), and together they produce 40 percent of Sweden's electricity, equal to hydropower, with the rest coming mainly from biofuels and wind power. These power plants have never had any serious accidents and fewer incidents altogether than almost any other industry would routinely expect. Nobody has died from the kärnkraft (notwithstanding a few fatal industrial accidents at the power plants, unrelated to the power source). Nobody has choked on the polluting exhaust because there isn't any. The plants can run at about 90 percent capacity on average over the year, producing electricity reliably around the clock.[9] Sweden's economy has thrived in the kärnkraft era, with cheap electricity for industrial, commercial, and residential use. Sweden enjoys a relatively high level of energy use per person and stays warm during its cold northern winters.

The largest of the four kärnkraft sites is at Ringhals, on Sweden's west coast. On just 150 acres (1/4 square mile), it can produce up to 4 gigawatts of electricity, 24/7. It averages 24 terawatt-hours of electricity generated in a normal year.[10] Like Ringhals, Sweden's other two kärnkraft sites, the Forsmark and Oskarshamn plants, generate large amounts of electricity cleanly and quietly in lush, scenic, and tranquil locations on the coast of the Baltic Sea.

What if the same electricity were produced by other means?[11] Replacing the Ringhals plant with one fueled by coal would require almost 11 million tons of coal each year—a train more than 1,300 miles long, producing 2 million tons of toxic solid waste (ash, mercury, and more), including radioactive components,[12] and spewing huge clouds of particulates into the air—enough to kill about 700 Swedes each year.[13] It would also produce about 22 million tons of CO_2 to accelerate climate change.[14] Coal miners would die in accidents and suffer from black-lung disease. Landscapes would be leveled by strip mines.

Getting the same electricity from oil would be much more expensive and somewhat less polluting than coal—but

Figure 6. Sweden's Ringhals power plant. *Photo*: Courtesy of Annika Örnborg / Vattenfall.

only somewhat. To replace Ringhals, every year 40 million barrels of oil would be pumped and transported, the size of twenty supertankers, with risks of massive spills at sea, from pipelines, or from trains. Burning the oil would release particulates, though less than coal, and pour 17 million tons of CO_2 into the atmosphere. Money paid to oil-exporting countries might prop up some unsavory dictatorships and fuel armed conflicts.

Methane (natural gas) would produce electricity more cleanly in terms of particulates in the air. But it would still add about half as much CO_2 as coal, compared with nearly zero for kärnkraft.[15] Producing methane often involves "fracking" by injecting toxic chemicals under pressure into rock formations—chemicals that sometimes end up in local water supplies. Methane is explosive, and occasionally buildings or whole city blocks blow up in lethal explosions. Getting large quantities of methane to places like Sweden often requires liquefying it at very low temperatures for transport in expensive tankers. Alternatively, it could be piped in from Russia, which is geopolitically problematic.

Wind turbines could also make the same amount of electricity in a year, but replacing Ringhals would actually require about three times the power capacity because wind is variable and produces about a third of its peak capacity on average.[16] To produce the energy generated at Ringhals would require twice the wind capacity of the wind farm planned for Markbygden in Sweden, which is to be one of the largest wind farms in Europe. A Ringhals-equivalent double-Markbygden wind farm, the state of the art, would entail 2,500 turbines, 650 feet high, spread over 400 square miles. But although such a wind farm would produce the same energy as Ringhals over a year, it would be variable, sometimes far higher than demand and sometimes far lower. There is currently no generally available and economically practical way to store that energy for when it's needed.[17]

Solar panels also could not replace Ringhals. Even if solar power were not intermittent, about 20 GW of solar power would be needed to produce the same amount of electricity as Ringhals's 4 GW of kärnkraft, because the sun only shines during daytime and only on some days at that. In a dark northern country such as Sweden, very little solar power can be produced in winter, when energy needs are highest. The scale of a 20 GW solar farm is hard to imagine, covering 40–100 square miles.[18] Imagine driving down a highway at 65 mph, with solar cells stretched out for a mile to the right of you and a mile to the left. It would take you about half an hour before you got to the end of the solar farm. After serving their technical life span of about twenty-five years, the panels would have to be disposed of, including their toxic materials, and replaced with new ones. By contrast, the amount of annual waste from the Ringhals kärnkraft plant—the equivalent of about six railroad boxcars or fifteen shipping containers—can be buried in one safe location at the end of the plant's fifty-year life, and the disposal costs have been included in the electricity price, so the money is there when it's needed in the future.[19]

Given these choices, it's not hard to see why Sweden opted for kärnkraft. Even before it considered climate change, Sweden decided kärnkraft was superior to coal, oil, methane, wind, or solar power—or suffering energy scarcity as the country shivered through cold winters. The extra benefit was that Sweden became the most successful country in history at expanding low-carbon electricity generation and leading the way in addressing climate change.

Sweden was not alone, though. France, Belgium, and Switzerland did more or less the same thing. France built

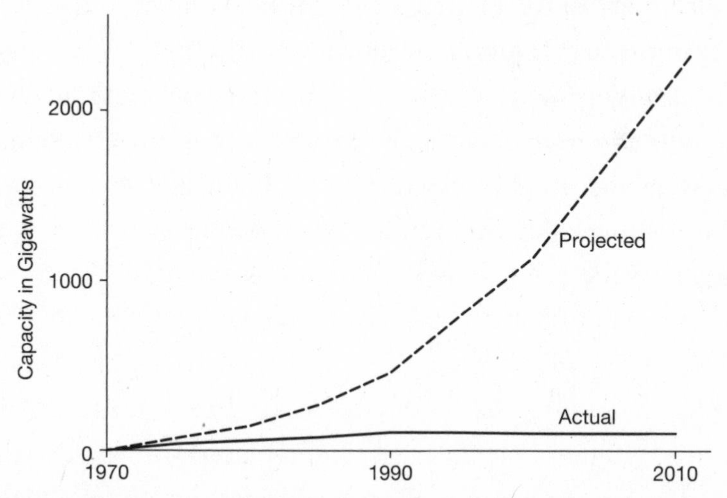

Figure 7. Projected growth of US kärnkraft capacity, US government, 1974, versus actual growth. *Source*: Projection from US Atomic Energy Commission. *Proposed Environmental Impact Statement on the Liquid Metal Fast Breeder Reactor (WASH-1535)*(1974). Actual-growth data from Harold A. Feiveson, "A Skeptic's View of Nuclear Energy," *Dædalus* (Fall 2009): 63.

fifty-six kärnkraft units in twenty years, has 70 percent lower carbon emissions per person than the United States, and enjoys the cheapest electricity in Europe.[20] The United States started down the same path but abruptly stopped just as Sweden was taking off.

Yet even today, kärnkraft provides one-fifth of all the electricity in the United States and two-thirds of its "clean" carbon-free electricity. Like in Sweden, nobody has died (except in occasional industrial accidents unrelated to the power source), almost no carbon has been emitted,[21] and hundreds of thousands of lives have been saved compared with burning coal (which still produces a substantial amount of electricity in the United States). Almost a hundred kärnkraft units are scattered around the United States, quietly and cleanly providing power around the clock without the oil spills and train crashes, the gas explosions, the coal mine disasters, the lethal air pollution, and all the rest.

If you haven't figured it out yet, *kärnkraft* is Swedish for "nuclear power."

CHAPTER THREE

What Germany Did

SWEDEN'S NEIGHBOR GERMANY has taken a very different path. Both are northern European, industrialized countries with successful economies. The GDP per person is almost the same. Sweden uses one-third *more* energy per person than Germany. Yet Germany emits about twice as much carbon pollution per person. Why is that?

Germany has received a tremendous amount of favorable publicity for its green *Energiewende* (energy transition) policy of installing large capacities of renewables—mostly wind and solar power. But how does that approach work out in terms of the world's need for rapid decarbonization?

In the past decade, Germany has roughly doubled its production of energy from renewables, an impressive accomplishment. In 2016 renewables made up more than a

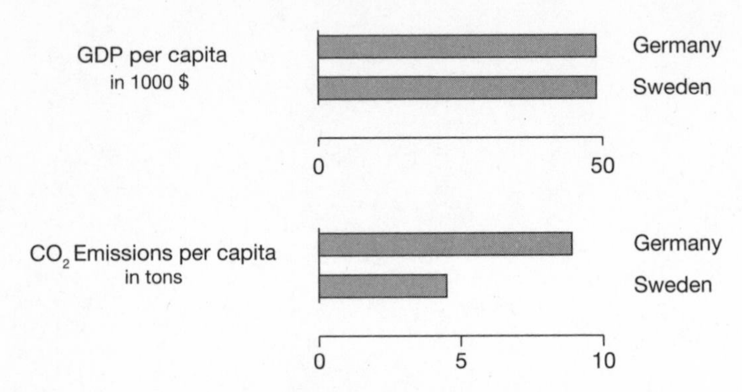

Figure 8. German versus Swedish emissions. *Sources*: GDP: World Bank (PPP), 2016; CO_2: Carbon Dioxide Information Analysis Center, 2014.

quarter of electricity production and almost 15 percent of total energy production.

But here's the catch: While it doubled renewables, Germany cut nuclear power by roughly an equivalent amount. It just substituted one carbon-free source for another, and CO_2 emissions did not really decrease at all. In fact, they have gone up slightly in recent years. And this will continue in coming years because, after the 2011 Fukushima accident in Japan, Germany is phasing out its remaining nuclear power in the next few years. The decade during which we desperately need to be rapidly decarbonizing will be a lost decade for Germany.

German energy remains dominated by fossil fuels, specifically coal, and not just coal but an especially dirty CO_2-heavy type of soft coal called lignite. In Germany's electricity production, lignite alone supplies almost a quarter, and all

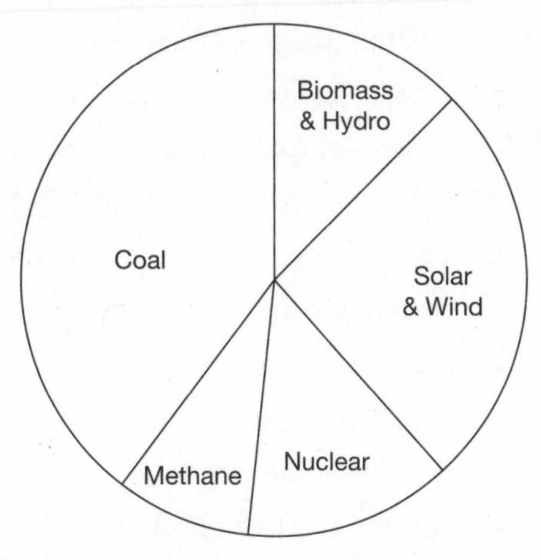

Figure 9. German electricity fuel mix, 2017. *Data source*: Fraunhofer ISE Energy Charts.

coal supplies 40 percent. Renewables are 29 percent and growing, but nuclear power is 13 percent and shrinking.[1] The coal burns on. Germany's greenhouse gas emissions remain around a billion tons a year.[2] If Germany had used its new renewables to replace coal, instead of to replace nuclear power, the CO_2 emissions picture would be quite different.

Germany's three biggest coal-fired power plants all burn lignite coal. One of them, the Jänschwalde power plant, is located about 400 miles south of the Ringhals nuclear power plant in Sweden. At Jänschwalde coal from nearby strip mines is brought to the plant in trains and burned to boil water and run steam turbines. The electricity goes onto Germany's electric grid, which is integrated with northern Europe's grid. In a good year, the plant produces almost as much electricity as Ringhals (20 TWh/year),[3] but it does so in a dramatically different way.

Jänschwalde burns vast quantities of coal and emits vast quantities of CO_2. On average it burns at least 50,000 tons of dirty coal per day.[4] If you loaded that coal into a single coal train, it would stretch 5 miles long. If you made elephants out of coal and burned them to boil water, you would march about 10,000 elephants into the maw of the great machine every day. The next day, another five miles of coal train, or another 10,000 elephants. The CO_2, more

Figure 10. German lignite coal strip mine with the Jänschwalde plant in the background. *Source*: Courtesy of Hanno Böck.

than 60,000 tons daily, is not captured and sequestered; it is dumped into the atmosphere.[5] A fair estimate of the deaths caused by the air pollution from this one coal plant is about 650 people per year, with 6,000 additional people suffering serious illness.[6] Germany has two other similar lignite coal plants and a number of smaller ones.

On the World Wildlife Federation's list of the most polluting power plants in Europe, measured by CO_2 emissions per unit of electricity produced, Jänschwalde is number four. Six of the top ten are German.[7] In 2016 the German state secretary for economic affairs and energy predicted that Germany would continue to burn lignite coal until after 2040.[8]

The company that owns the Jänschwalde plant sees coal as integral to Energiewende: "The expansion of renewable energies and the phasing out of nuclear energy by 2022 are the two main objectives of the 'Energiewende' (energy transition). Our lignite power plants accompany this process. On the one hand, they offer a reliable round-the-clock supply and on the other hand, they are flexible and capable of adjusting their own production to the current available renewables which have the right of way in the transmission grid."[9] In 2015 a top executive of the then owner of Jänschwalde discussed coal's role in supplying Germany's projected need for 20 to 50 GW of baseload generation: "Nuclear power plants will be decommissioned in a few years. This means that only lignite and hard coal fired power plants will be left for base load generation. In Germany we currently have around [20 GW] installed capacity in lignite-fired power plants. I am convinced that we have to retain this capacity on the long-term.... Lignite will continue to play a considerable part in the future energy mix in the next four to five decades."[10]

Wind and Solar

Germany has adopted a policy to switch its economy to renewables. If the country first cut fossil fuels rapidly and then transitioned from nuclear power to renewables, this policy might be defensible. Instead, Germany is running in place by reducing clean nuclear power as it increases clean wind and solar.

Wind and solar power are wonderful, but they start from a very small share of our energy systems, and, to date, they have not been a fast way to replace fossil fuels. They are inherently diffuse in contrast to concentrated energy sources, including both coal and especially nuclear power. And they are variable and uncertain.

Consider the largest solar power facility in Europe, the Solarpark Meuro in Germany. It covers about 500 acres on a former lignite coal strip-mining site. This is an excellent use for a former coal mine, and the 166 megawatts (MW) of solar power are essentially carbon-free.[11]

The problem is twofold—scale and timing. Even though Solarpark Meuro is the largest solar site in Europe, you would need twenty of them at peak capacity to equal one

Figure 11. Solar farms such as this one in France, or Solarpark Meuro in Germany, require large land areas. *Photo*: Mike Fouque / Shutterstock.

Ringhals or Jänschwalde. Worse still, that peak is produced only during the best season, the best weather, and the best time of day. Every night, every winter, and every cloudy day, production is closer to zero. As a rule of thumb, nuclear power produces at 80–90 percent of capacity on average over the year, coal at around 50–60 percent, and solar cells around 20 percent. So to get the actual electricity production of one Jänschwalde would require about seventy Meuro solar farms. As solar prices drop, Germany might one day build those seventy huge solar parks, and might even use them to attempt to "replace" Jänschwalde instead of replacing nuclear power plants, but even then the power would not be available at the times it is needed, so backup power from fossil fuels would still be needed. Clearly, such an approach is not a formula for rapid decarbonization.

Wind power is more important than solar for Germany. Many wind turbines have been built in recent years, mostly on land rather than offshore because it typically costs less. Germany does not have massive wind farms that could begin to compare with power plants such as Jänschwalde, but it could build these along the lines of Europe's largest wind project, Fantanele-Cogealac, in nearby Romania, built in 2008–2012.

The Romanian wind farm covers 2,700 acres, about three times the size of Central Park in New York. The 240 wind turbines are each as tall as a fifty-story skyscraper,

Figure 12. A small part of the Fantanele-Cogealac wind farm in Romania, Europe's largest. *Photo*: Courtesy of ČEZ.

and the diameter of their rotors is thirty stories tall.[12] So how much electricity does all that steel and concrete generate? The total peak capacity is 600 MW, about one-sixth of Jänschwalde. But wind is only somewhat more reliable than sunshine, and production in 2013 came in at below 25 percent of peak capacity.[13] To equal the production of one Jänschwalde would take about 13 wind farms of this size, equipped with modern, state-of-the-art wind turbines.[14] Even then, that production would not happen when needed, but variably, sometimes too much and sometimes too little.

In Germany wind has been somewhat unreliable. From 2015 to 2016, Germany added 10 percent more wind capacity but generated less than 1 percent more electricity from wind, because the wind did not blow as much that year.

In 2017, as Germany integrated solar and wind onto the grid—still just 7 percent and 12 percent of generation respectively, very far from 100 percent—intermittency began to affect the grid significantly. As one German grid operator put it, "We have a lot of stress on the grid." Although solar and wind generation frequently dropped to near zero at some times, it surged to exceed demand at other times, even with the greatest practical short-term cutback of coal and nuclear production. More than a hundred times in 2017, sometimes for more than a day at a time, electricity prices went negative. Grid operators paid large consumers as much as 6 cents/kWh (almost 10 cents on one occasion) to take power to avoid overloading the grid.[15] Until someone invents a supercheap battery with huge capacity—no time soon, it appears—this instability from relying on wind and solar power will only worsen.

You might think that unevenness in supply would even out across countries, given a large integrated regional grid. "The wind is always blowing or the sun shining somewhere," according to this line of thinking. But our analysis of the eleven top wind-producing countries and five top solar-producing countries in Europe in 2013 showed a period of forty-eight hours in which wind produced at only 6 percent of capacity across the whole continent, an entire month in which solar produced at only 3 percent, and a week in which all wind and solar combined across Europe produced at less than 10 percent of their capacity.[16] During that week, powering the European continent with wind and solar power would require installing either massive redundant capacity that would sit idle most of the year or an extensive fossil-fuel infrastructure able to supply almost all of the electricity demand in Europe during the periods when renewables dropped out.

In addition to the intermittency problems, Germany's massive rollout of renewable power does not match the speed of Sweden's nuclear power rollout decades earlier. During the very fastest period of adding renewable electricity in Germany (2014–2015), the expansion relative to GDP size was less than a third of the rate of Sweden's expansion

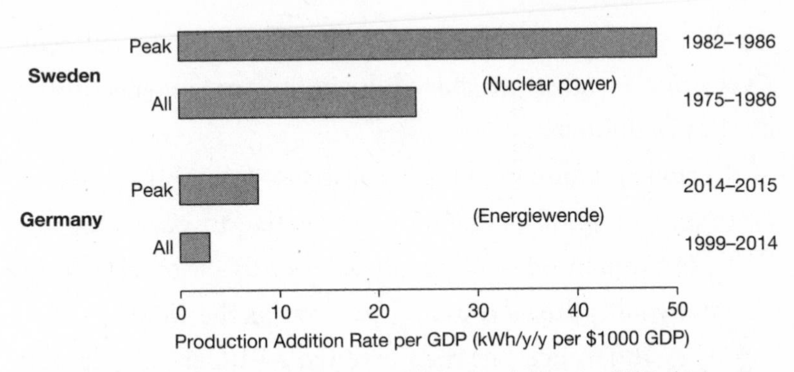

Figure 13. Rates of clean energy additions relative to GDP during peak periods, Germany versus Sweden. *Source*: Generation data for all countries are from British Petroleum, *BP Statistical Review of World Energy* (2017). Population and GDP data in constant 2005 dollars are from World Bank Databank, *World Development Indicators* (2018). Gross nuclear generation values from BP are reduced 4.1 percent to obtain net values.

of clean energy during its decade of nuclear power rollout and one-seventh of the peak rate in Sweden. This means that what the world might achieve in, say, twenty years using Sweden's kärnkraft model would take more than a hundred years using Germany's Energiewende model. Those are a hundred years the world doesn't have.

Several conclusions jump out from Germany's experience. First, in order to rapidly decarbonize, a country must shut its largest, dirtiest coal-burning power plants and must mobilize any available resources such as new wind and solar capacity toward that end. Second, adding renewables to the mix is an excellent thing to do, but these sources alone cannot scale up quickly enough to rapidly decarbonize, especially given the intermittency problem. Third, phasing out nuclear power is in direct competition to phasing out coal. Every nuclear gigawatt closed down is a fossil gigawatt that continues to burn.

Germany's failure to learn these lessons explains why it continues to spew twice the CO_2 relative to economic activity as Sweden does. When it comes to a green economy, Germany talks the talk, but Sweden walks the walk.

A key difference between Sweden and Germany is that Sweden uses *both* nuclear power and renewables—equal parts hydropower and nuclear power with a growing wind component—along with biomass cogeneration plants for district heating. This approach recognizes that fixing climate change requires all the tools available and does not afford us the luxury of picking only those we like best, since they can't succeed in isolation.

We can call the Swedish combination of nuclear power plus renewables "nuables." It adds up to a solution that works. Nuables are doable.

HALF MEASURES

Current efforts that move in the right direction—energy conservation, the growth of renewables, and shifting from coal to methane—cannot alone achieve the rapid decarbonization we need.

CHAPTER FOUR

More Energy, Not Less

SWEDEN USES A lot of energy per person. It is among the top ten consumers of energy per person in the world.[1] It is a modern, urbanized country with industry, commerce, transportation, and very cold winters. Yet it has successfully decarbonized its electricity grid. Energy conservation and efficiency are fine. They move in the right direction to slow climate change, especially in the less efficient places such as America. But Sweden's high level of energy use per person shows that using less energy is not the key to decarbonizing.

Actually, all the world's rich, industrialized countries use a lot of energy, which currently comes overwhelmingly from fossil fuels. It's true that some, like the United States and Australia, use more per person, whereas others, like Japan and Britain, use less. Americans drive bigger cars than Japanese do, and drive more rather than take trains, and they live in bigger homes that are kept warmer in winter and cooler in summer. Americans consequently emit more carbon pollution than do people in more energy-efficient countries. But if America and Australia used energy the way Japan and Britain do, it still would not solve the climate crisis.[2]

Sure, technological progress has helped industrialized countries, including the United States, to become more energy efficient, halting the growth of carbon emissions and even bringing them down modestly while economic growth continues. LED lightbulbs are replacing compact fluorescents, which replaced incandescents. Gas mileage for cars keeps improving. Smart thermostats deliver heat to our homes when and where needed. Airplane engines use less fuel.

But flattening out emissions through efficiency is not what we need. We need rapid decarbonization. No amount of energy conservation can deliver that; only an extreme economic depression would reduce energy consumption that much, or the asteroid that destroys most of humanity. The answer is not less energy, but cleaner energy.[3] We want a future of *Star Trek*, not *Blade Runner*.

Yes, we need everything that helps, and energy conservation helps. So by all means recycle, ride your bike, and become a vegetarian. But do not imagine that these actions alone will solve the problem.[4] Unplugging your phone charger when not in use will save the amount of energy used in one second of driving your car.[5] Some personal actions such as lowering the thermostat make a difference on a personal scale, but a lot of personal actions that demand effort and attention are basically just feel-good distractions. In the extreme case, Americans drive gas-guzzling SUVs to the recycling center and imagine they are helping solve the problem. The idea that personal behavioral changes would add up to solve global environmental problems has been popular among environmentalists for decades—President Jimmy Carter put on a sweater and set the thermostat down in the White House in the 1970s—but this approach has in fact had virtually no impact on the world's carbon emissions.

The "rebound effect" also reduces the impact of energy-efficiency measures, although the amounts are uncertain.[6] When cars and other equipment become more energy efficient, there is an incentive to use them more. For example, the US automobile fleet has become more and more fuel efficient, but at the same time Americans have driven more cars for longer distances, with the result that gasoline consumption is higher now than in the 1990s, even though cars are much more efficient.[7] In the coming years, self-driving cars will make driving even more efficient but might also tempt people to commute for longer distances owing to the ease of the commute. Overall, energy efficiency has made a positive difference in slowing or even modestly reducing energy use in richer countries, but not very rapidly or dramatically. Of course, energy efficiency does allow more and better uses of energy to improve life without *increasing* the total amount used. But that's not decarbonization.

A bigger problem is that as fast as richer countries can wring energy efficiency out of their economies, poorer countries' energy use is growing rapidly. Energy use per person in poorer countries is about one-tenth of that in richer ones, so there is a lot of potential for growth. And that growth is a good thing because it is lifting billions of people out of poverty and improving the lives of billions more. But it means that decarbonization requires both replacing existing fossil fuels and meeting new demand from carbon-free sources.

Many more people live in the poorer countries than the richer ones, and they want to become richer. They want more energy, and they have a moral right to it. Energy helps lift people out of abject poverty. As China has rapidly increased its energy use (and its carbon pollution), it has brought hundreds of millions of people out of poverty and into a more comfortable lifestyle. Today's rich industrialized countries did the same thing over the past two centuries, and it's their carbon that today is overloading the atmosphere. To tell the poor countries that there is no more room for their carbon, and they must stay poor, would be morally wrong and impractical.

The population of India, well over a billion people, wants lights, clean water, refrigerated food, air conditioners, cars, houses, hospitals, and so forth. The demand for air-conditioning alone, just in India, is projected to add demand for 150 gigawatts of electricity by 2030, just as Chinese citizens over the past fifteen years have added more than 200 gigawatts from their 200 million new air conditioners.[8] Worldwide, led by India and China, demand for air-conditioning is projected to grow from about 700 gigawatts now to 2,300 by 2050, the period in which the world must decarbonize.[9] Given that a nuclear or coal power plant typically has at most a few gigawatts of capacity (and renewable installations far less), these numbers are staggering. But the massive expansion of air-conditioning will improve lives and health greatly in the hot, humid countries of the world.

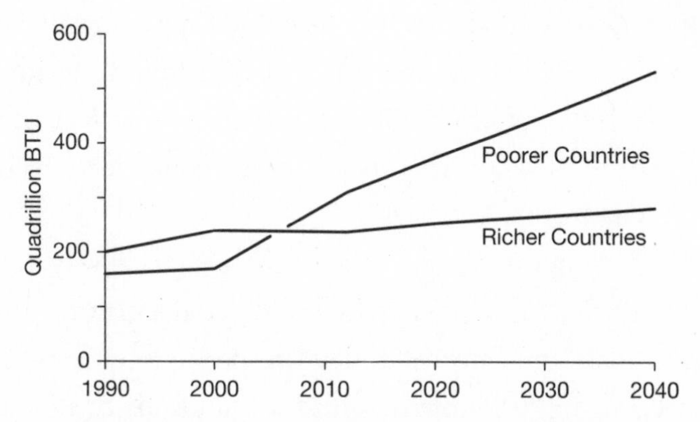

Figure 14. Projected energy consumption for Organization for Economic Cooperation and Development and non-OECD countries. *Data source*: US Energy Information Agency, *International Energy Outlook, 2017*.

In fact, all of Asia, Africa, and Latin America wants the same energy-intensive development that China has achieved. Projected worldwide consumption of all types of energy thirty years from now is about 50 percent higher than today—growth that absolutely overwhelms any energy-efficiency efforts that we successfully implement. Most of that growth will occur in poorer countries.[10]

More than a *billion* people in the world lack access to electricity, a quarter of them in India, but the number is slowly shrinking as the grid reaches more communities and keeps up with world population growth. Every year almost 100 million people get access to electricity for the first time.[11] So every year we have to find new electricity supplies for these new consumers, about 25 million people in India and 75 million more around the world. Even though we get more and more goods and services for each kilowatt-hour, the total consumption of those kilowatt-hours worldwide will roughly double by 2050. Almost three-quarters of this electricity will be added in the poorer countries.[12] Where will it come from?

If the cheapest, fastest, and most practical way to get more electricity is to build a lot of coal-burning power plants—as it typically is today in poor countries—then that's what those countries will do. Hundreds of new coal plants are being built right now. India has been putting in large solar farms, but that's only slowing down the growth of coal power in India. In 2014 when Greenpeace installed a $400,000 solar-powered microgrid for an Indian village, electricity use in the village went up, and batteries could not provide enough storage. Greenpeace put up posters urging residents to save energy by not using rice cookers, water and space heaters, irons, and air coolers. After residents protested, "We want real electricity, not fake electricity!" the village was connected to the grid—a grid powered by coal at one-third the price of solar electricity.[13]

In 2017 the Indian government stated that, despite plans to increase both renewables and nuclear power, "the reality of India's energy sector is that around three-quarters of our power comes from coal powered plants and this scenario will not change significantly over the coming decades.

Figure 15. Indians without electricity grind sugarcane into feed by hand, 2015. *Photo*: Courtesy of Ibrahim Malik.

Thus, it is important that India increases its domestic coal production."[14]

If we don't want the huge increases in CO_2 that come with those coal plants in developing countries, we need something better that is equally cheap, fast, and practical. What we can't do is tell countries like India not to increase their energy use. We can't tell a billion people that they won't get electricity because we already burned up all the carbon the planet can afford.[15] The world needs not less energy but different energy.

To add 100 million people to the grid per year, each using just a tenth of the electricity of the average US consumer,[16] would require new electricity production of 130 TWh/year. That's about five times the production of Ringhals (Sweden's largest nuclear power station; see Chapter 2) or more than six massive coal-burning plants like Germany's Jänschwalde (see Chapter 3). Every year. Just for the new customers. Of course, the existing customers in poorer countries desperately want more electricity as well. If not from fossil fuels, where will it come from?

Population

A common misconception among people in richer countries such as American environmentalists is that world population growth is a driver of climate change. The truth is that rising energy use per person, not the rising number of people, drives carbon emissions. Imagine that the poorest half of humanity ceased to exist, leaving the West and China. In that world, population growth would be about zero, but carbon emissions would be almost as high as they are today. Poor people don't use much energy.

The huge rise in carbon emissions in past decades has come from the richer parts of the world, the very places where population growth has stopped and even turned negative in some places. This is not a coincidence. Population growth follows an *S* curve as a country's income rises. In the beginning, people are poor, and both birth and death rates are high. As incomes begin to rise, death rates fall with access to basic health care, and population grows. As incomes rise further, birth rates fall because parents are more confident in their babies' survival and because women gain access to education and contraception. At the end of this "demographic transition," both birth and death rates are low, more of the population is older, and incomes are higher.

Higher incomes, and higher energy use, are the cure for population growth.

The world as a whole is well along in this transition. In the past fifty years, the world fertility rate (births per woman) fell from about 5 to 2.5, where a bit over 2 is the steady-state replacement rate (zero population growth). India's fell from 6 to 2.4 and China's from 6 to 1.6[17]—yet China's carbon emissions have skyrocketed because of its higher incomes. Europe, Russia, and Japan now have negative population growth but continue to dump huge amounts of carbon into the atmosphere because of their high energy use per person. Focusing on "overpopulation" is a distraction at best and at worst a rather racist view that blames problems created by the rich countries on people in the poor countries. Fears in the 1960s that overpopulation would lead to disaster—in particular, Paul Ehrlich's prediction in the 1968 book *The Population Bomb* that hundreds of millions would starve in the 1970s and humanity would break down by the 1980s—did not come true.[18]

The way to further reduce population growth is to raise incomes in poor countries more quickly. However, this means using more energy. More energy use means more carbon emissions, unless we decarbonize. Up until now, only one carbon-free energy source has proven able to scale up very quickly and—in the right conditions—affordably. That source is nuclear power.

CHAPTER FIVE

100 Percent Renewables?

"WE DON'T NEED nuclear power" goes the current conventional thinking among environmentalists. "We can build 100 percent renewables." The demand to generate all the world's energy needs from renewable sources—hydroelectric power, wind power, and solar power—has become a mantra for much of the climate-action movement.[1] Tell your city and state to provide 100 percent renewables. Power your corporation with 100 percent renewables. As a slogan, it's catchy and simple. And it moves in the right direction. Every time a solar farm replaces a fossil power plant, that's a victory.

But it doesn't add up to a complete solution, certainly not in the upcoming decades in which climate solutions must be found.[2] Over the past decade, the world has spent $2 trillion on wind and solar power but has seen almost no progress toward decarbonization.[3]

To start with, new renewables alone do not scale up fast enough for the rapid decarbonization that we need, even if they were not variable and uncertain (which, as we shall see, is a serious issue). We need renewables to make their contribution, but we also need all other tools in the box to achieve rapid decarbonization. We need "nuables"—nuclear power *plus* renewables—both scaling up as fast as possible. But nuclear power can scale up faster than renewables can. Fundamentally, this is because the energy in nuclear fuel is millions of times more concentrated than wind or solar power.

To assess this question of speed, consider the historical experience of countries' rollout of nuclear power compared with the rates of rolling out renewables in recent years. In assessing alternatives for decarbonizing the world, a key measure is this: How much carbon-free energy was a country able to add per year, relative to population or GDP, during a peak decade of rollout? By this measure, all the recent increases in renewables fall far short of what was accomplished decades ago in rolling out nuclear power in various countries—with Sweden leading the list. For instance, each year during the peak decade of 2005–2015, German wind and solar power combined added about 120 kWh/year per person. California added about 70. By comparison, Sweden in its peak decade added over 600 kWh/person each year, and France added 450.[4]

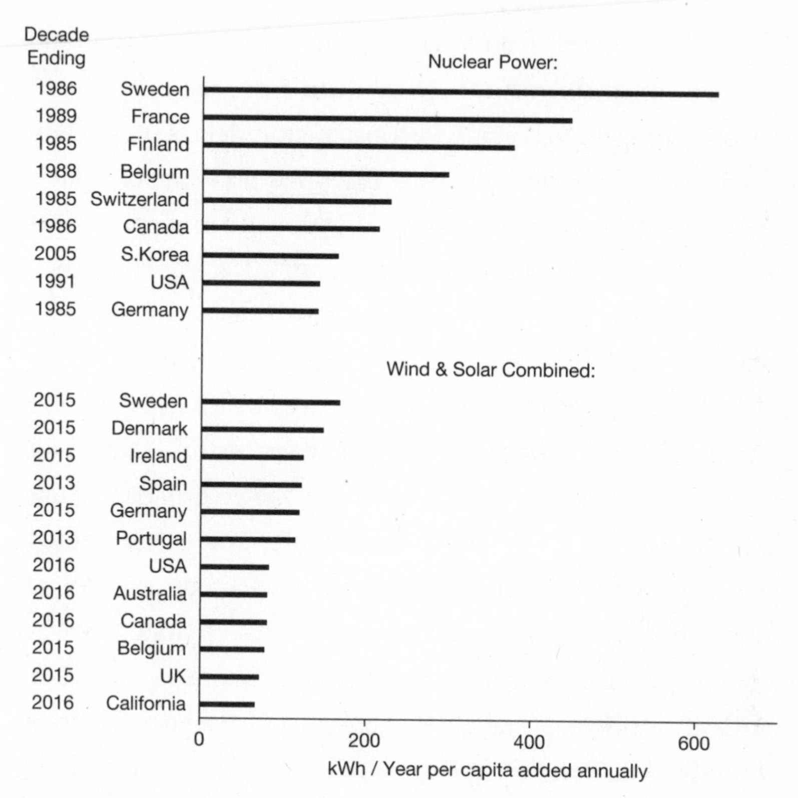

Figure 16. Carbon-free electricity added per person per year during peak decade. *Data source*: British Petroleum, *BP Statistical Review of World Energy*; World Bank Databank, *World Development Indicators*.

The point we made in Chapter 3 about Germany's Energiewende policy applies globally: What the world already knows how to do in ten to twenty years using nuclear power would take more than a century using renewables alone. The *story* of using only renewables seems compelling, but the *scale* does not work to rapidly decarbonize the world.

Nuclear power, even today with its many challenges, still adds power quickly. Although Finland's new EPR has faced massive budget overruns and construction delays, it will add clean electricity faster, over that extended time line of delays, than the world record for doing so with wind and solar combined. Upon connection with the grid in 2018, the new Finnish reactor was to generate about as much electricity annually as all the wind turbines Denmark has built since 1990.[5]

India has committed to an impressive goal of installing 100 GW of solar power by 2022. Yet, even if successful, which is far from certain, this effort would only slow the growth of coal use in India. For one thing, the 100 GW of intermittent solar would produce only the electricity equivalent to about 25 GW of round-the-clock coal (or nuclear power). And that new power would be installed over five years, so it adds about 5 GW a year, modestly more than a single Ringhals plant each year. In terms of production, that new solar power could generate something like 40 terawatt-hours per year. But just the new customers on India's grid each year require almost that much electricity (see the previous chapter).[6] With existing customers wanting a lot more electricity too, even this ambitious Indian solar plan will not reduce coal use. Similarly in other developing countries, the growth of renewables can at best keep up with new demand, not displace existing coal power.

China, despite being the world's leader in renewable energy, continues to burn mountains of coal. Coal accounted for 72 percent of China's electricity production in 2015,

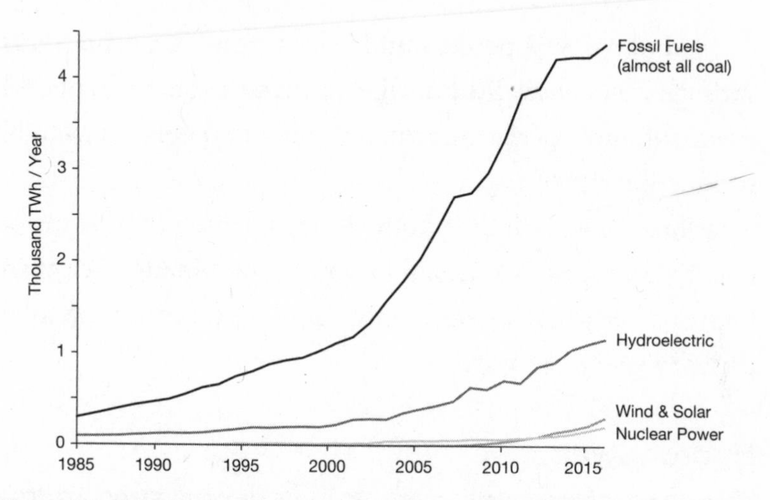

Figure 17. China's electricity generation by fuel, 1985–2016. *Data source*: US Energy Information Agency.

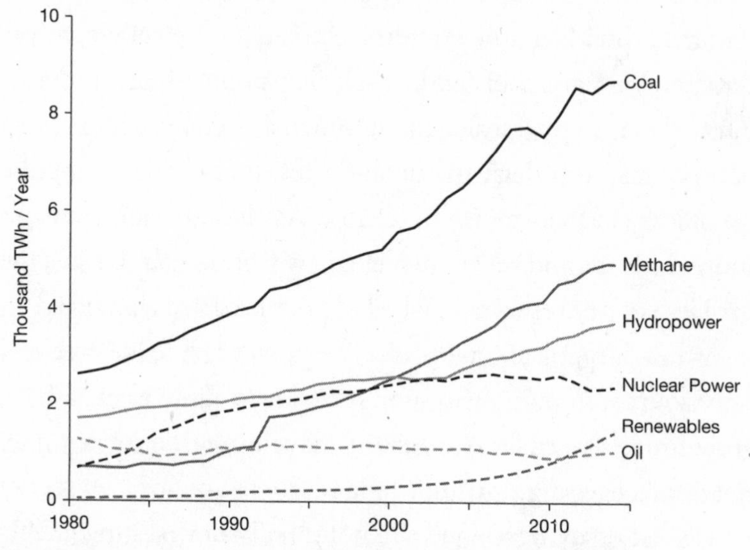

Figure 18. World electricity generation by fuel, 1980–2014, in terawatt-hours. Renewables refers to wind, solar, biomass, and geothermal power. *Data source*: World Bank.

with wind below 3 percent and solar below 1 percent. Plans through 2040 show little decline in the actual *amount* of coal used, although its *share* declines to about 50 percent as total production rises.[7]

Worldwide, including both electricity and other forms of energy, all types of renewable power combined other than hydropower today provide only about 3 percent of the total primary energy supply.[8]

Hydropower

Two-thirds of renewable energy worldwide comes from hydroelectric power. Hydropower is great from a carbon-mitigating point of view, but most of the prime sites have already been dammed. Building a new hydroelectric dam typically involves flooding vast areas of land, displacing populations, and embarking on a large-scale, capital-intensive construction project. Because of a lack of suitable sites, hydro does not scale up quickly in the way we need now. As climate change causes more frequent and severe droughts, hydropower also faces the problem of dry reservoirs, which do not produce electricity.

While large-scale hydroelectric power is a fossil-free energy source, in many ways far superior to fossil alternatives, sometimes the rush to hydro in developing countries may have far-reaching and unintended consequences. A recent *Economist* essay described the potentially devastating effects of the many hydroelectric dams being developed on the Mekong River in Southeast Asia, one of the most biodiverse ecosystems on the planet and the foundation for 15 percent of global rice production.[9] In 2018, one of the more than fifty hydroelectric dams under construction in Laos burst and killed scores of people.[10] As more and more dams are developed, fisheries, agriculture, and biodiversity will suffer, the extent of which we have no good way of estimating. This again highlights the benefits of a portfolio approach to clean energy—simultaneously building solar, wind, hydro, and nuclear power to avoid having to rely on environmentally damaging or risky projects.

Wind

Wind power produces electricity at reasonably low cost. However, it needs a lot of area, and the windiest places are generally far from the cities where electricity is consumed. This means large, expensive upgrades to the electric grid, such as Texas recently undertook. Offshore wind could help to solve this problem, since wind farms offshore could be close to coastal cities—and offshore winds are generally steadier than those onshore—but prime locations are somewhat limited. For example, California's seafloor drops off rather steeply compared with Britain's. Sweden's largest offshore wind farm, at Lillgrund near the bridge to Denmark, was built in 2007 in a good location, in shallow water close to shore, but produces electricity at about 11 cents/kWh compared with about 4 cents for methane and is losing money.[11] Globally, a recent analysis estimates offshore wind costs at 11 cents/kWh versus 3–6 cents for onshore

Figure 19. Offshore wind is a key part of renewables expansion, but still expensive. Sweden's Lillgrund wind farm, 2007. *Photo*: Mariusz Paździora via Wikimedia Common (CC BY-SA 3.0).

wind.[12] And offshore wind farms generally are supposed to last for twenty-five years, but there is some evidence that shorter life spans than expected may add to the ultimate price.[13]

Nonetheless, offshore wind costs have dropped significantly in recent years. In 2017 a large offshore wind project in Britain was priced at about 8 cents/kWh, half the price of just a few years earlier.[14] Other prices are higher, however. The first offshore wind farm in the United States—Deepwater Wind off Block Island, Rhode Island, completed in December 2016—produces electricity at almost 24 cents/kWh. A newly approved offshore wind farm near Long Island, not yet built, will come in at 16 cents, which is still more than double the utility's average electricity rate.[15]

And wind is variable. Not only does it blow more at some times of day than others, but in some years more than others. As noted in Chapter 3, Germany's wind power produced about 10 percent less electricity in 2016 than in 2015 relative to installed capacity.

Replacing coal power plants with nuclear power plants is straightforward. Both are "baseload" power, available around the clock. By contrast, trying to replace coal with wind or other renewables is complicated by the variable and uncertain nature of renewables' production, which must be balanced by either hydropower or methane (which can ramp generation up and down quickly). In the extreme case, which sometimes occurs, a rush of wind power onto the grid on a windy day forces steam from a nearby coal plant to be vented (and hence wasted) rather than used to run a turbine and generate electricity. It is simply not feasible for the coal plant to power up and down to match the ups and downs in the wind. New methane plants can do this more feasibly (at some cost), but this problem means that the rollout of renewables often goes hand in hand with a rollout of methane power to balance the load. More often, rather than vent steam from a coal plant, electric grids "curtail" wind or solar power, wasting their potential output rather than letting it on the grid. This has particularly affected China's massive wind farms, where curtailment has recently caused a reduction of about 20 percent in wind-power generation in the country.[16]

Solar

Solar power is abundant but starts from barely more than 1 percent of world electricity supply today. Like wind, it has an important role to play but also has limits.

If you sit at your computer in the United States some afternoon and look at the map of real-time energy generation by source in Europe, you may be struck by a startling observation. In Europe it is evening, and—although lights are on, people are moving, and heating or air-conditioning is cranking—the solar generation across the continent is zero.[17] Solar power is not just variable like wind but totally unavailable for major parts of every day. Hydropower in some countries can compensate for this daily fluctuation, as it can for ups and downs in wind, but in northern countries solar power drops drastically for months at a time. Solar production is near zero in winter, just when lighting and heating needs peak.

The intermittency of solar power means not only too little at times but too much at other times. The grid cannot accommodate big surges during peak solar production. California gets more than 10 percent of its electricity from utility-scale solar power and another 4 percent from rooftop solar. As with Chinese wind power, California has had to force curtailments on its solar production—15 percent of the time in 2015 growing to 30 percent in early 2017. Because solar is decentralized and hard to control, especially for rooftops, California still regularly ends up with too much production and, to avoid overloading the grid, pays Arizona to take some of it.

This is called "negative pricing," frequently at a substantial price of 2.5 cents/kWh.[18] (Germany has a similar negative pricing problem, as mentioned in Chapter 3.)

California renewables are mandated by the state government to grow from 25 percent of total electricity to 50 percent by 2030. A new 2018 regulation requires that every new home in the state have solar power.[19] At the same time, California is removing the 10 percent of electricity still generated by baseload 24/7 nuclear power. Hydropower production is becoming unstable as climate changes, because drought one year becomes deluge the next. So the problem of intermittency will get much worse in the coming decade. California's electricity, already about 50 percent more expensive than the national average, will likely increase further in price.

Solar advocates celebrate the large amounts of solar power being installed these days, but the added *capacity* does not translate to production in the same way other sources do. Headlines that compare installation of new solar capacity with new coal capacity are not as positive as they sound—first, because solar produces far less power for each GW of capacity and, second, because installing *any* amount of new coal power is a step in the wrong direction when we need a power source to *replace* coal quickly.[20]

Solar costs have dropped steeply in the past few years. If not for intermittency, solar power could compete with fossil fuels in many places. The International Energy Agency (IEA)'s analysis of levelized costs of energy, for generating sources entering service in 2022, lists solar farms at 7.4 cents/kWh and methane at 5.4 cents, before subsidies.[21] The solar costs continue to drop.

But solar *plus* long-term battery storage, to solve intermittency, is still impractical and quite expensive. Recently, news reports heralded an agreement in Tucson, Arizona, to provide solar power plus battery storage for just 4.5 cents/kWh, a price competitive with coal or gas generation. On further inspection, however, the cost was heavily subsidized by the city, and the battery storage was good for only about fifteen minutes of storage, just to smooth out the grid from the shock of crashing production each afternoon when the sun goes down. Fifteen minutes after the sun sets, the system generates electricity from methane or coal.[22]

The cost of rooftop (or "distributed") solar is much higher than "utility-scale" solar power in large farms. By one analysis, utility-scale solar in a favorable location such as the Southwest United States, and with no consideration of the costs of intermittency, can cost as little as 5 cents/kWh. Residential rooftop solar in the same analysis costs 19–32 cents,[23] although costs continue to drop. Solar cells on rooftops are an iconic image of cool renewables but not really practical as a grid-scale solution to replacing fossil fuels.

The definitive new book on solar power, Varun Sivaram's *Taming the Sun,* warns that our current path will not lead to a solar-powered one-third share of world electricity by 2050, needed to reach climate targets.[24] As more and more solar is added to the grid, its value falls even though the price of the solar cells keeps declining. This is because solar power produces cheap electricity only at times when there is already a surplus of electricity—more and more of a surplus as solar's share of the total grows.

Although cheap solar cells resulted from massive subsidies both at the producing end (especially in China) and the consuming end (mostly Western countries), by now economies of scale, and at times oversupply, have made them cheap even without subsidies. Some new installations have

Figure 20. Europe by night: lights on but solar power off. National Aeronautics and Space Administration composite, 2014. *Photo*: NASA Goddard.

guaranteed prices as low as around 3 cents/kWh, lower than any other source.

When solar is just a few percent of the total, the grid absorbs it easily, and the cheap price is a blessing. But as solar grows to 10 or 20 percent of the total, this production dominates the grid when the sun is shining and suddenly leaves a huge gap when night, clouds, or seasons stop the sun shining. We have mentioned that electricity prices go negative when renewables production is high in Germany and California. Adding more solar cells means producing more at these times. Electricity at 3 cents/kWh is no bargain when electricity is being given away free. A cheap ice cream cone in a remote desert would be a bargain, but a thousand more would be worthless unless you had a freezer. You could neither eat them nor save them. Currently, we lack affordable, long-term, grid-scale electricity storage—the equivalent of the ice cream freezer.

Because of problems such as this, solar power in the leading European solar countries—Italy, Greece, Germany, and Spain—topped out at 5–10 percent of total electricity and saw no growth at all from 2014 to 2016.[25] California has pushed past 15 percent of the total, but at a cost. On a sunny day, almost half of California's electricity comes from solar power, and methane is only 10 percent. When the sun goes down, methane and imports from Nevada and Arizona (methane, coal, and nuclear power) fire up to fill that gap, about one-third of total demand. In the morning the same massive shift happens in reverse.[26] To accommodate those surges, the grid needs backup fossil-fuel plants and transmission lines at the ready, as well as upgrades to help the grid cope with variability. These costs add up. Turning methane plants on and off is inefficient and wears out equipment. Germany is spending $20 billion on grid upgrades and expensive backup plants, and its neighbors complain that German fluctuations destabilize their own grids.[27]

The hidden costs of integrating solar onto the grid add about 50 percent to the stated cost of solar power, according to Sivaram.[28] This is in addition to subsidies from producing and consuming countries. None of these costs are included when cheap solar power is discussed. In California solar generators get paid for their electricity production whether it's needed or not.[29] As a result, California has the most expensive electricity in the country, rising from 35 percent above the national average to 60 percent above just from 2011 to 2017 as renewables expanded. German consumers pay one-quarter of their high bills to support renewables. In fact, everywhere that renewables have expanded most, electricity prices have risen.[30] Solar farms also do have environmental impacts (though certainly far fewer than coal). They eat up large amounts of land because sunshine, although potent in the aggregate, is spread out across Earth. Ramping up solar production would mean paving over many square miles at a time, often consisting of productive farmland or undeveloped nature, with steel and silicon. This loss of cropland and habitat runs counter to other climate-change goals. The neighbors do not always approve of the change, and solar projects can be tied up in court for years. At the back end of the cycle, solar cells last only about twenty-five years, after which they must be recycled, which is a large-scale, dirty, toxic operation often carried out by children in very poor countries with few safeguards. Unlike for nuclear power, the costs of decommissioning solar farms are not usually included in the price.

Batteries

Because wind and sun both come and go, a grid dominated by renewables would depend on cheap energy storage, such as new breakthrough battery technologies. However, these do not yet exist and may not for years. Today's best consumer batteries, such as Tesla's Powerwall unit, are too expensive for grid-scale storage. (The very largest of such batteries, such as Tesla's famous battery in Australia, can provide some short-term stabilization to the grid at a price competitive with other methods such as methane peaker plants. Using such batteries to hold renewable power long term is still not affordable.) The potentially cheapest large-scale batteries—flow batteries using large tanks of liquid—are more promising but still too expensive. So far, lithium-ion batteries are the most economical and proven. Large-scale battery installations are dropping in cost as the industry matures.[31] Used on a grid scale, battery storage adds about 30 cents/kWh to the cost of electricity, whereas for "behind the meter" commercial and residential use the cost is 85 cents to $1.27/kWh.[32] (Recall that the average US electricity price is now 10 cents/kWh.)

For the very cheapest solar installation available—a thin-film utility-scale solar farm in the Southwest United States—a recent comprehensive analysis found that the cost of adding just ten hours of storage would nearly double the cost of electricity.[33] This makes solar plus storage more expensive than fossil fuel and impractical for developing countries, much less for locations where the sun fails to shine for more than ten hours at a time.

In a world that currently uses about 68 TWh/day of electricity,[34] the investment in batteries to store just one day's worth of production would exceed $20 trillion, which is about a quarter of all economic activity in the world in a year. That is just the storage cost, added to the cost of producing electricity. Nor is it clear that one day of storage would suffice; as we noted in Chapter 3 regarding Germany's Energiewende ambitions, there are periods of a week when solar and wind combined drop below 10 percent of capacity for all of Europe.

Production of batteries to power the world on renewables would be mind-boggling in terms of not only cost but also production capacity and supply of raw materials. The world's current annual production of lithium-ion batteries, like those in electric vehicles and Tesla Powerwalls, would power the world's electricity needs for about forty-five seconds. Tesla's "gigafactory" in Nevada will double this rate of world total lithium battery production when it reaches full production.[35] If you built a new gigafactory every year (the first one took five years), it would take sixty years to cumulatively produce enough batteries to hold just one day's storage of the world's electricity. However, since batteries have a limited life of charging cycles, even the most high-end batteries typically do not last longer than fifteen years when operated on a daily load cycle, so more decades would be needed to replace the dead batteries from the early decades. And, of course, the "one day's storage" at today's electricity consumption rate would not meet the need in sixty years when world electricity consumption will probably have doubled. Solving climate change requires much faster decarbonization than that.

Bill Gates, who has invested $1 billion in renewables, states, "There's no battery technology that's even close to allowing us to take all of our energy from renewables and be able to use battery storage in order to deal not only with the 24-hour cycle but also with long periods of time where it's cloudy and you don't have sun or you don't have wind."[36] As the latest report from Lazard, the authoritative source on energy costs, puts it, "Although alternative energy is increasingly cost-competitive and storage technology holds great promise, alternative energy systems alone will not be capable of meeting the base-load generation needs of a developed economy for the foreseeable future."[37]

All these problems might be overcome with enough time. Perhaps later in the century, if there were a huge breakthrough in the cost of batteries, we could find the US northeastern seaboard powered by massive offshore wind farms. But it would be very irresponsible to depend, for humanity's future, on solutions that we hope will appear decades from now and that depend on technological breakthroughs that have not yet occurred—when we have methods available that are already proven to work. Cheap batteries or fusion power might save us. The asteroid might miss us altogether. But these are not the chances a responsible person takes. Wind and solar power have a growing and vital role to play in replacing fossil fuels, but starting from just 5 percent of world electricity supply, that role alone does not scale up fast enough to make the math work to say "we don't need nuclear power."

Making Solutions Add Up

The "Solutions Project" and Stanford professor Mark Jacobson caused a huge stir recently with their claim that the United States[38] and the world[39] could be both cheaply and reliably powered by 100 percent renewables by midcentury. A subsequent article in the same journal by a distinguished group of experts, including other Stanford professors, rebuts these claims.[40] In the US case, the 100 percent–renewable scenario depended heavily on huge increases in hydropower that do not appear to be feasible. Because hydropower can balance out the variability of wind and solar power, such as by opening up more water flow at night when solar cells are offline, this assumption of vastly increased hydropower propped up the whole scenario. It doesn't add up. Jacobson's response to his critics was to sue them and the National Academy of Sciences in court for $10 million—an extremely irregular way to address an academic dispute in a scientific journal. (He later dropped the lawsuit, leaving legal costs for the defendants to pay and a chilling effect on the research community.)

What does add up is an important and growing role for hydro, wind, and solar power in the coming decades. The faster these energy sources are deployed, the easier will be the job of rapidly decarbonizing. Partnered with nuclear power in a "nuables" solution, they are a key part of fixing climate change.

The mistake is thinking that those steps in the right direction will add up and solve climate change alone. They won't. Bolstering renewables to reach 50 percent of the world's growing electricity production would be a great step in the coming decades. But "100 percent renewables" is a slogan that distracts from the work at hand, which is the decarbonization of the world.

An example of this distraction factor is the Climate Simulator published on the *New York Times* website in 2017.[41] Based on the MIT model described in Chapter 1, it brilliantly steps the reader through the math to show the need for immediate, massive decreases in CO_2 emissions if the world is to stay within limits. However, the graphic then claims (without any evidence) that, good news, such cuts "may be possible" with wind, solar power, and energy efficiency. These solutions, however, cannot achieve precisely what the model itself shows to be necessary—a rapid decrease in emissions. A more useful simulator would go on to let the reader try out various technologies, to see that without a major expansion of nuclear power, the targets simply cannot be met.

Beyond just distracting from solutions that add up, the 100 percent–renewables idea has been used repeatedly as a rationale to shut down existing zero-carbon nuclear power plants with the idea that they can be "replaced" with renewables. But as we build out renewables, we absolutely must use them to replace fossil fuels, not carbon-free nuclear power. After the last fossil power plant closes, and we have only nuclear power and renewables, then we could talk about whether to replace the nuclear power capacity with renewables if it proved practical and beneficial. This is what some Swedes hope to do in a few decades, although the published science shows that it would actually be better environmentally and economically to keep Sweden's nuclear power plants running.[42]

Figure 21. "Replacing nuclear with renewables" does not decrease carbon emissions. *Graphic*: Vaclav Volrab / Shutterstock.

So far, we have discussed wind and solar power as additions to the electrical grid, but one version of the all-renewables argument holds that communities can use renewables to break free of the grid altogether. Chapter 4 discussed the failure of this approach in an experiment in India. It has also been tried recently for a small village in southern Sweden. A German utility powered the village with a local microgrid supplied with locally produced solar and wind electricity, without reliance on the national electricity grid. But in practice, the system needed to draw more than 80 percent of its electricity supply from the national grid.[43] The utility bluntly noted, "If you look at this when it's very cold outside, the wind is rarely blowing and it's also dark, so the solar cells are not producing. That's the way it is, and everyone knows that."[44] For each unit of electricity the village installation does manage to produce from its own sources—which include solar cells, wind turbines, batteries, and a biodiesel backup generator—the carbon emissions are higher than for electricity imported from the national grid.[45]

The 2017 book and website *Drawdown* lists eighty "solutions" that move in the right direction, casting them as "the most comprehensive plan ever proposed to reverse global warming." The solutions range from obvious ones such as rooftop solar panels to indirect ones such as expanding girls' education, as well as future technologies that do not yet exist, such as artificial leaves and hydrogen-boron fusion. Adding up the solutions in a "plausible scenario," which they describe as "reasonable yet optimistic," the authors find the solutions actually would not achieve the needed drawdown.[46]

One of the *Drawdown* solutions is nuclear power (number twenty on the list of eighty), which the authors assume will grow by 2030 and still provide 12 percent of the world's electricity by 2050. Given the need to "do everything we can," as the authors put it, one might expect strong support for expanding nuclear power's role. Instead, the editor has added a special "Editor's Note" unique to the nuclear power solution, stating that while almost all the other solutions are "no-regrets" actions with many beneficial effects, nuclear power is a "regrets solution" because of the negative effects. He lists fourteen names of places where nuclear power problems have occurred, such as "Browns Ferry"—evidently a reference to a 1975 fire that did not cause a meltdown, human casualties, or release of radiation.[47] What he does not claim is that we can solve climate change without this "regrets solution."

The *Drawdown* approach is far more comprehensive and sophisticated than a simple "all we need is 100 percent renewables" line. At the same time, it suffers from a similar problem, which is to focus on steps that move in the right direction without examining what feasible measures can actually solve the problem of climate change.

The promotion of 100 percent renewables also contributes to the skewing of public opinion away from nuclear power. After a well-funded, decades-long global fear campaign against nuclear power, people are anxious about it, and renewables seem to offer a comfortable alternative to combat climate change without confronting those anxieties. In China and India, a 2015 public opinion poll shows that about half the public supports the development of more renewables, about a quarter supports the expansion of fossil fuels, and less than 10 percent supports the expansion of nuclear power.[48] In the West, too, publics support clean, cool options but do not evaluate whether they actually solve the problem.

When a company, university, or town declares that it has achieved "100 percent renewable" electricity, that statement is not true. It should say *net* power of 100 percent, with a lot of extra clean energy sold to the grid part of the time (often when it's least needed) and a lot of dirty energy of equivalent amount bought from the grid at other times. As we have seen, this is a far cry from not needing dirty energy. True reliance on 100 percent renewables would mean disconnecting from the grid without relying on backup fossil-fueled generators. Almost nobody has done this because it is not practical or affordable. And currently it is no more practical for a country than for a company.

Renewables are an important *part* of the solution to climate change. Costs for wind power and, especially, solar power are dropping dramatically in recent years.[49] In places where they can be feasibly and economically added to the grid, they can help displace fossil fuels. (This is far more practical when they are a small part of the total on the grid. For example, it makes more sense to focus on adding renewables in China, where three-quarters of the electricity comes from coal and only about 5 percent from wind and solar, than in Germany and California, where renewables already provide about a third of electricity.) So, by all means, let's build renewables, but let's keep our attention focused on what needs to be done in the next ten to twenty years to rapidly decarbonize the world and not fall into the delusion that 100 percent renewables is the solution.

CHAPTER SIX

Methane Is Still Fossil

LONG AGO, THE oil industry began calling methane gas, produced in conjunction with oil drilling, "natural gas." This distinguished it from synthetic gas produced in a dirty process from coal a century ago, but "natural" makes it sound environmentally friendly as well. It isn't. All the fossil fuels are molecules made of hydrogen and carbon compressed underground over the millennia and then dug up and burned. Methane (CH_4) burns more "cleanly" than coal and oil, with fewer toxic by-products, such as the small particles in coal smoke that cause lung cancer and emphysema. (Commercial natural gas also contains less than 10 percent molecules other than methane, which we disregard here.)[1]

Methane creates less CO_2 than coal when burned, per unit of energy output—about half. Much of the progress that industrialized countries have made in flattening and modestly reducing carbon emissions in recent years has come about by substituting methane for coal. This is especially so in the United States, where revolutionary methods known as "fracking" (hydraulic fracturing by injecting liquid into rock formations to release gas) have made methane cheaper than coal. This cheap methane is the main factor behind the decline in coal use—not a liberal "war on coal"—and also behind the economic troubles of the nuclear power industry. Neither fuel can compete with US methane, nor can renewables without mandates and subsidies. In fact, methane power plants that can ramp up and down quickly make a handy complement to renewables with variable production, so the addition of renewables to a grid often really means adding renewables *and* methane.

Internationally, though less dramatically, other countries are also turning to methane. Russia produces large amounts and exports it to Germany and the rest of Europe, though with geopolitical tensions sometimes getting in the way. (Key pipelines go through war-torn Ukraine.) And many countries have started to import and use liquefied natural gas (LNG). It is not as cheap as methane piped from the fracked well to the consumer, because the gas must be cryogenically cooled to liquefy it and then transported in that state in tankers. But still it's an increasingly popular fuel.

Cheaper and cleaner than coal...What could be the problem? There are two main ones from a climate perspective—in addition to the local problems of water contamination near fracking sites.

First, methane may produce only half the CO_2 of coal, but that's still a massive amount of CO_2. As consumption rises, so does the CO_2. Yes, a growing amount of methane burned is better than a growing amount of coal, but it still runs counter to the needed rapid decarbonization. Worse yet, investments in methane infrastructure such as pipelines and LNG terminals are multibillion-dollar projects that will take decades to pay back and therefore commit us to a fossil-fuel economy well into the future.

Figure 22. Using methane gas to replace coal and balance renewables means building extensive new fossil-fuel infrastructure, from fracking wells to LNG terminals to pipeline networks. This carrier delivers LNG to Japan to replace nuclear power, 2012. *Photo:* とまりん^^ via Wikimedia Commons (CC Attribution-Share Alike 2.1 Japan).

Then, too, the switch from coal to methane can happen only once, so the gains in limiting carbon emissions are only temporary. While transitioning from coal to gas, a country will make notable progress toward its goals for reducing CO_2, but as that transition finishes, suddenly emissions will stop dropping.

The transition itself can be difficult and expensive, for something that still leaves an economy dependent on fossil fuel. In the United States, cheap fracking makes methane an economical fuel, but only with investment in large-scale pipeline infrastructure. In China the supply of methane is much smaller and the switch from coal more challenging. In 2017 China tried to force a more rapid switch from coal to methane, with authorities dismantling and carting away coal-fired boilers from schools, homes, and businesses. Then winter arrived earlier and colder than usual, and methane shortages quickly developed. Elementary schools were without heat, and a major chemical company could not fulfill its contracts—leading to a world shortage of spandex, used in clothing. By December the government turned a large coal plant back on and shut down big chemical factories for four months to conserve methane for schools and homes.[2] However, the strategy, in combination with a phaseout of polluting motorbikes and dirty cookstoves, did clean up the air in Chinese cities considerably in the winter of 2018.[3]

The second major problem with methane is that it leaks, especially at the well but also all along the pipeline system.[4]

Unburned methane is a potent greenhouse gas but breaks down in the atmosphere much faster than CO_2, lasting only decades rather than hundreds of years. In a period of a couple of decades, a ton of unburned methane has more than 80 times more warming effect than a ton of CO_2, and even over a century as it breaks down it is 25 times more powerful. Some experts, though not all, think that the entire shift from coal to methane is not really a gain in climate terms.[5] Given that much more CO_2 is being released from the burning of gas, unburned methane is less important in contributing to climate change, but it's a serious issue. The decades when today's methane leaks are having their strong warming effect are the decades when the climate is destabilizing and we need to be acting quickly to reverse the trend. Unburned methane moves us rapidly in the other direction, and the problem is only getting worse as methane becomes more popular as a fuel.

In Los Angeles in 2015, a massive leak at an underground methane-gas storage facility led to health problems and the evacuation of a whole neighborhood. During the four months it took to control the leak, around 100,000 tons of methane went into the atmosphere (equivalent to several months of CO_2 emissions from the whole Los Angeles Basin).

Cows and other cattle are another major source of greenhouse gases, contributing about 7 billion tons of CO_2 equivalent each year, or about 15 percent of the human

Figure 23. Los Angeles methane leak seen in infrared photo, 2016.
Photo: Courtesy of Environmental Defense Fund.

contribution to climate change. About half of the cattle contribution is unburned methane gas, primarily from cow digestion (mostly burps). The problem can be reduced with changes in feed, one of a series of changes in agriculture and land use that should be better propagated and funded in order to reduce climate impacts.[6] (Substituting chicken or pork for beef in our diets or going vegan would also greatly reduce both unburned methane and energy use.)

The concentration of methane in the atmosphere rose ten times more quickly in 2006–2016 than in the prior decade, the fastest rise in recent decades. This increase is largely unexplained, perhaps related to rice production and probably not caused much by fossil-fuel use, but it threatens to set back progress toward reducing global warming.[7]

Less important from a climate perspective, but worth noting, methane gas causes deadly explosions from time to time when it leaks and then combusts. In October 2017, an explosion at a liquefied natural gas facility in Ghana killed seven and injured more than a hundred.[8] In 2010 a gas pipeline near San Francisco exploded in a residential neighborhood, sending a wall of flame hundreds of feet in the air, destroying thirty-five houses and killing eight people. In South Korea in 1995, a gas explosion killed more than 100, most of them teenagers. In 1937, a methane explosion at a school in Texas killed almost 300 children.

Methane has been marketed as natural, clean, efficient, and cheap. But its role in addressing climate change is limited. It does have a role, but we should not consider it a solution.

FACING FEARS

Fixing climate change requires setting aside habitual fears and understanding that carbon-free nuclear power is vastly safer than today's dominant energy source, coal.

CHAPTER SEVEN

Safest Energy Ever

IN 2011 A massive earthquake and tsunami hit the east coast of Japan, a bit north of the Fukushima district. In the fishing town of Onagawa, a nearly 50-foot wall of water destroyed everything and left hundreds of survivors homeless. They fled to the safest and most secure location they knew about, the Onagawa nuclear power plant. There they sheltered and received food and blankets. That's right—sometimes the safest place to be in an epic natural disaster is your local nuclear power plant.[1]

Onagawa's three reactors, dating from the 1980s and '90s, had a capacity of more than 2 GW of electricity, more than half that of Ringhals in Sweden. Like in Sweden, the reactors had operated for decades without a major accident. On the fateful day in 2011, the plant's 46-foot-high seawall prevented major flooding, and despite a lot of earthquake-caused cracks in the building, the reactors all shut down normally and without incident. No radiation was released; nobody was hurt.

Down the coast a bit, and twice as far from the epicenter, the Fukushima Daiichi nuclear power plant did not fare as well. The reactors there depended on backup diesel generators to keep coolant flowing, and although there were many of these backups, they were all flooded by the massive tsunami, which overwhelmed the plant's seawall, whose height was less than 20 feet. (Note that the problem was not the reactor design but the unforgiveable decision to locate all the backup generators together in a location vulnerable to flooding with too small a seawall.) Over several days, among various problems in several reactors, the core of one reactor overheated into a radioactive mess and released hydrogen gas that exploded, breaching the containment structure. Radioactivity leaked into the surrounding environment and ocean. A panicky, botched, and almost entirely unnecessary evacuation ensued, displacing hundreds of thousands of residents in the area.

How much harm that radiation did is a subject of controversy. But if you trust science, you should believe the conclusions reached by the extremely thorough studies done by several UN agencies, including the World Health Organization. These experts all reached the same answer to the question of how many people were killed, either directly through high radiation exposure or likely to die later through elevated rates of cancer in the population. That answer: approximately zero. (We will discuss this conclusion in more depth at the end of this chapter.)

The health risks from the Fukushima reactors (even when employing the most conservative analysis possible) were in fact so low that in retrospect, the optimal response would have been to not evacuate anyone.[2] The unnecessary evacuation of hundreds of thousands of people may have caused about 50 deaths among patients moved from hospitals,[3] and as many as 1,600 deaths in the longer term, owing to elevated mortality from causes such as obesity, diabetes, smoking, and suicide among psychologically stressed evacuees.[4] It is hard to know how many critically ill patients would have died absent the evacuation and how many psychological stresses should be attributed to the evacuation rather than the earthquake/tsunami disaster itself. Mortality rates returned to normal levels after about a year,[5] although more than 100,000 evacuees remained.

Elsewhere in Japan, by contrast, the earthquake and tsunami themselves killed about 18,000 people, injured many more, and caused hundreds of billions of dollars in damage. It was, after all, the largest earthquake in recorded history in Japan and the third largest ever worldwide. Yet, just five years later, the trauma of this epic disaster seems all but lost in the world's collective memory, and discussion of the event focuses on the Fukushima "nuclear disaster." In truth,

Figure 24. Tsunami damage near Fukushima, 2011, unrelated to the subsequent nuclear power accident. *Photo*: Toshifumi Kitamura/AFP/Getty Images.

there was no nuclear disaster at Fukushima. There was a natural disaster of biblical proportions, a small consequence of which was a very expensive and disruptive but nonlethal industrial accident at the Fukushima power plant, followed by an unnecessary and botched evacuation.

What happened next, however, killed thousands of people. Japan and Germany panicked (there is no better term for it) about nuclear power and shut down perfectly good, safe nuclear power plants. Japan closed fifty-four reactors and Germany eight more. Almost all remain closed six years later. These were inevitably replaced mostly with fossil fuels, including a lot of coal, and those fossils polluted the air with their particulates and toxins, increasing cancer and emphysema in the population.[6] Although an exact estimate is difficult to make, the deaths from this switch to fossil fuels were certainly in the thousands each year, or easily more than 10,000 over six years.[7]

So here is an accounting of the toll of the 2011 earthquake/tsunami: earthquake and tsunami disaster, 18,000 killed; nuclear power "disaster," nobody killed; botched evacuation, perhaps on the order of 1,000 killed; slow-motion disaster of replacing clean nuclear power with dirty fossils, more than 10,000 killed. Radiation rarely kills anyone, but fear of radiation kills a lot of people.[8]

Three Mile Island and Chernobyl

What about the other famous nuclear power accidents?

Three Mile Island was the most serious nuclear power accident in the United States. In 1979 a reactor partially melted down when it overheated, but the containment structure prevented radiation from affecting the surroundings. It was expensive but harmless. Unfortunately, the accident happened just as a fictional nuclear power disaster movie, *The China Syndrome*, starring Jane Fonda, was captivating audiences. The public saw the reactor accident as proof that nuclear power was a disaster in waiting, as the movie had implied. The fact that the containment structure worked, keeping radiation from leaking, was lost in the wave of panic.

In 1986 in Chernobyl, Ukraine (then part of the Soviet Union), a nuclear power accident struck a reactor that did not have a containment structure. The accident, caused by bad design and a series of operator errors, resulted in significant release of radiation into the environment. The Soviet government tried to keep it secret, and the radiation spread across northern Europe before the government finally admitted the problem. This government response meant that lifesaving actions such as providing iodine pills to local residents did not happen.

How many people died? A few dozen on the scene died, mostly first responders who fought the reactor fire in very high-radiation conditions. The UN experts did their careful studies of the radiation effects and concluded that, most likely, up to "several thousand" people could eventually die from cancer as a result of the radiation exposure, although the increase among a large population would be so small as to be "very difficult to detect."[9]

The Chernobyl reactor was eventually encased in a concrete sarcophagus, and an exclusion zone of 1,000 square miles around the plant was evacuated and still has restricted access. Several decades later, scientists studying the exclusion zone are seeing one of the healthiest ecosystems in Europe. The absence of people has done wonders for the animals and plants there, whereas the higher levels of radiation do not seem to have affected them as much, notwithstanding a high initial impact in the most contaminated locations.[10] The point is not that everything was okay after the Chernobyl accident—it wasn't—but rather that the world's worst nuclear power accident in history was far less deadly than many recent earthquakes, hurricanes, industrial accidents, or epidemics have been. As has been shown since, the vast majority of evacuations from the Chernobyl area had no scientific justification.[11]

So this, then, is the safety record of nuclear power over more than fifty years, encompassing more than 16,000 reactor-years:[12] one serious fatal accident in the USSR with possibly, over time, up to 4,000 deaths; one Japanese "disaster" that caused no deaths; and one American accident that destroyed an expensive facility but otherwise just generated vast quantities of fearful hype. In the United States, nuclear power continues to produce about one-fifth of the nation's electric supply and has never killed anyone.

How do other energy sources stack up in terms of safety? Start with coal. As long as coal predominates in world energy supply, and even continues to be used in the (natural gas–oriented) United States, the role of nuclear power must be considered as a choice of nuclear power versus coal. Coal replaced some of the nuclear power capacity taken offline in Japan and Germany after 2011. In the United States, the two partly built reactors abandoned in South Carolina in 2017 had been slated to replace the state's coal plants, which will instead keep operating.[13] Environmental groups successfully blocked nuclear power construction in Ohio in the 1970s but stood by quietly as coal plants were built instead, and today that state still mainly uses coal for electricity generation. One Ohio nuclear reactor, 97 percent complete, was actually converted to burn coal after protests and problems stopped construction.[14] In Madison, Wisconsin, power is supplied today mainly from a large coal plant, because the nuclear power plant that would have supplied Madison was canceled under political protests decades ago. So nuclear power should be compared against coal as long as coal is still around in significant quantities.

As an order of magnitude approximation, coal kills at least a million people every year worldwide, mostly through particulate emissions that give people cancer and other

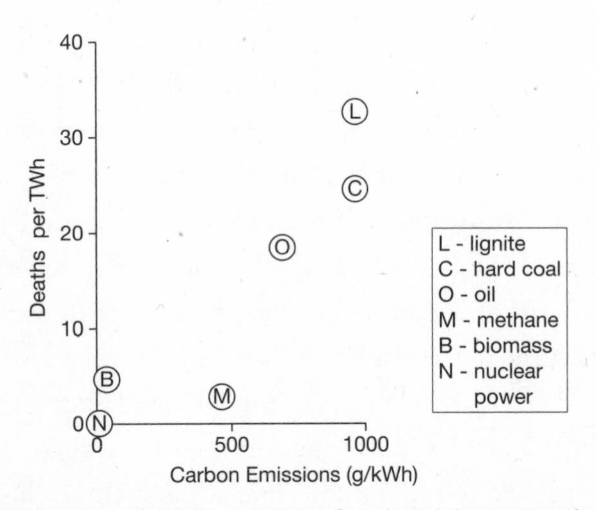

Figure 25. Deaths and CO_2 emissions for electricity generation in Europe, by fuel. *Source*: Anil Markandya and Paul Wilkinson "Electricity Generation and Health," *Lancet* 370 (2007): 981. Used by permission.

diseases. They die horrible, painful deaths, but mostly invisible ones, unlike the victims of spectacular events. Of the roughly 10,000 TWh of electricity produced by coal worldwide each year, about 3,000 are in the richer countries and 7,000 in the poorer countries.[15] Mortality effects have been estimated at 29 deaths/TWh in Europe and 77 in China,[16] which suggests an order of magnitude of 600,000 deaths a year just from coal use in generating electricity. (Other uses of coal such as for heating and industrial power are widespread and also deadly.)

In addition to its air-pollution effects, coal also has a terrible safety record, from legendary coal-mining accidents (still happening multiple times a year around the world) to toxic wastes around coal plants, generally located near poor communities. For example, in 1972, a failed dam sent a 30-foot wave of coal-waste sludge into sixteen towns in West Virginia, killing 125 people. In 2008 enough poisonous coal ash spilled at a Tennessee power plant to fill a football stadium almost a half mile high—more than the amount of oil spilled in the *Deepwater Horizon* accident.[17] Greenpeace demanded to know whether the authorities could have prevented the accident.[18] Why, yes. They could have approved the prototype breeder reactor about 20 miles away that the US Senate killed off in 1983.

Comparing deaths over the past fifty years from coal against deaths from nuclear power, including Chernobyl and Fukushima, the difference could hardly be more stark—tens of millions from coal and something like a few thousand from nuclear power. A large European study estimates coal deaths for each TWh of electricity at about 30, versus less than 0.1 for nuclear power—hundreds of times higher.[19] Over the decades, worldwide, the coal that nuclear power has displaced would have killed more than 2 million additional people—lives saved thanks to the extreme safety of nuclear power.[20] Cases of serious illness caused by coal and nuclear power are, in each case, about ten times higher than deaths, meaning hundreds of millions of cases of serious illness from coal pollution in recent decades.[21]

These safety comparisons are above and beyond the stark differences in CO_2 emissions. We would face a dilemma if the predominant energy source that caused vastly more global warming were much safer than the zero-carbon alternative. But it's exactly the opposite: the clean, zero-carbon energy source is also hundreds of times safer.

Other energy sources, although less dangerous than coal, still cannot touch the safety record of nuclear power.[22]

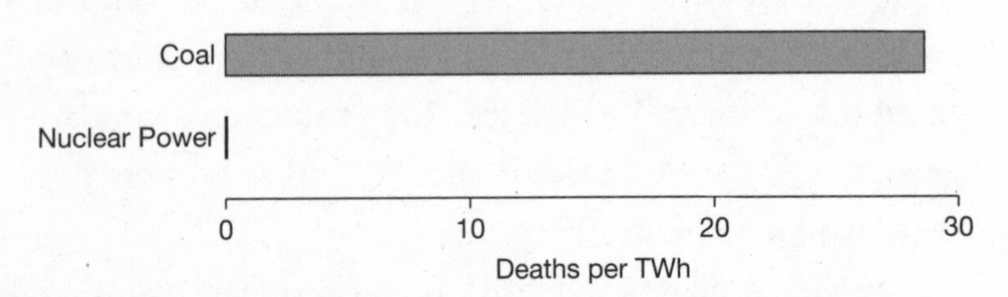

Figure 26. Deaths from nuclear power and coal in Europe per terawatt-hour of electricity. *Source*: Markandya and Wilkinson, "Electricity Generation and Health."

Methane gas blows up (see Chapter 6). Oil blows up and spills, as happened spectacularly in the *Deepwater Horizon* disaster in the Gulf of Mexico in 2010 (11 killed; 5 million barrels spilled). Oil trains are rolling bombs, as the town of Lac-Mégantic in Canada found out in 2013 when an oil train derailed, blew up, and leveled the town, killing 47 people.

Hydroelectric dams are far from safe. If a dam fails, it can flood communities downstream almost without warning. This happened in Banqiao, China, in 1975 and killed 170,000 people. Even in the United States, dam failures killed 2,200 people in 1889, 600 in 1928, and 238 in 1972, among others.[23] In 2017 alone, just in the United States, hydroelectric dam crises in California and Puerto Rico forced the evacuation of hundreds of thousands of people.[24]

Thinking About Radiation

Much of the public's concern about nuclear power comes down to a gut-level fear of radiation. Radiation is scary because it is invisible, potentially harmful, and associated in the popular imagination with nuclear weapons.[25] Radiation created Godzilla, the Fantastic Four, the Hulk, and Spider-Man.

Radiation is also a normal part of human existence and varies a lot in daily life. The unit that measures the impact of radiation that people receive is the millisievert (mSv). Background radiation in our daily lives averages around 3 mSv/

year, but it varies greatly. Smoking a pack of cigarettes a day adds about 9 mSv/year. Living in Denver or at a similar elevation adds about 2 mSv/year, while working on an airline crew on the New York–Tokyo route adds about 9 mSv/year because cosmic radiation is stronger at high altitude. Granite is radioactive, so living in a granite-rich location increases one's exposure compared with one near sedimentary soil. Spending all of one's time in Grand Central Terminal in New York, which is built from granite, would add about 5 mSv/year.[26]

The highest recorded natural background radiation is received by residents of Ramsar, Iran (where hot springs contain radium), with levels of more than 200 mSv/year. Although scientists and authorities have worried about such a high exposure, no evidence has emerged of negative health effects there.[27] There are many other, but less intense, natural "hot spots" around the world.[28]

Medical procedures also add to radiation exposure and overall account for about a third of the radiation to which humans are exposed (the other two-thirds being natural background radiation). A CT scan of the chest delivers about 7 mSv, in a concentrated time and place, but does not harm health.[29] The US Food and Drug Administration sets a limit of 50 mSv in a year for diagnostic procedures in adults. Levels used for treatment are much higher than for diagnostics. Radioactive iodine treatment of thyroid cancer delivers

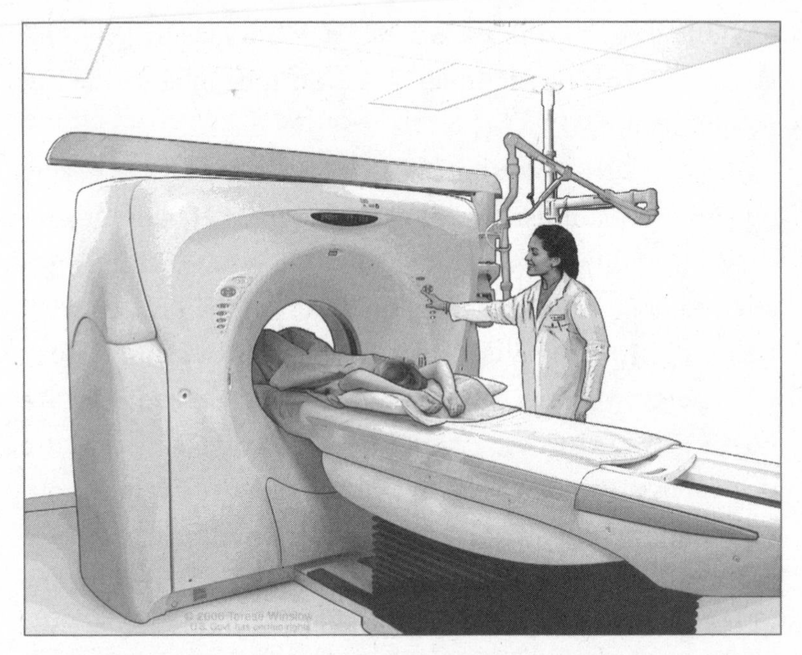

Figure 27. CT scans deliver radiation roughly comparable to levels near Fukushima. *Illustration*: Terese Winslow. Used by permission.

100,000 mSv to the thyroid and 200 mSv as a whole-body dose.[30]

Acute radiation effects kick in around 1,000 mSv, and at a rate of 60,000 mSv/year, the radiation kills half the people in a month.[31] So the people living in Ramsar, Iran, seem to do fine at 200 mSv/year, but you might not want to go a lot higher than that. In 2007 the International Commission on Radiological Protection (ICRP) recommended dose limits for occupational exposure of "no more than 50 mSv in any one year"[32]—the same as for medical diagnostics. This is also the exposure limit for those working in a US nuclear power plant.

In the 2011 Fukushima "disaster," radiation exposure outside the plant itself was low. The World Health Organization (WHO) estimated it at 10–50 mSv/year in the most affected areas and 1–10 mSv in the rest of the prefecture. It was orders of magnitude lower in other provinces. Among emergency workers, seventy-five individuals had effective doses of 100–200 mSv and twelve individuals received a dose above 200 mSv, and few health effects have been observed in them. Radiation doses beyond the first year were drastically lower.[33] Note that only a few people, and no one in the general population, even exceeded the ICRP recommended occupational dose limit of 50 mSv in a year.

Opponents of nuclear power rely on the "linear no-threshold" (LNT) assumption—that if a lot of radiation is bad, then a small amount is bad in the same proportion. According to this theory, exposure to 10 mSv in a thousand people will cause the same cancer harm as exposure to 1,000 mSv in ten people. The effects are extrapolated down from the very high levels of radiation received by victims of the atomic bombings in Hiroshima in 1945. This does not make any sense, given the similar health outcomes in places with different background radiation, such as Ramsar, Iran. It is a bit like saying that a hundred jumps from a one-foot cliff would have the same health impact as a single jump from a hundred-foot cliff. The human body handles little shocks over time, like many small jumps or small increases in radiation, better than it handles large shocks all at once.

There is even some scientific evidence that small amounts of radiation are actually beneficial.[34] But so far, nobody has a generally accepted formula, and nobody knows just *what* the threshold is, so the linear assumption remains the basis of policy in most places.[35] (The ICRP estimates that 200 mSv raises the risk of fatal cancer by 1 percent.)[36] One implication is that even a tiny additional dose of radiation in people such as those near Fukushima will translate, over a very large population, to increased cancer fatalities.

According to the LNT approach, exposure to the granite in the perfectly safe Grand Central Terminal—which as mentioned has slightly elevated radiation levels of about 5 mSv/year—is just a mild version of the Hiroshima bombing. With 750,000 visitors per day, if we assume each stays twenty minutes on average, Grand Central is causing two to three cancer deaths each year.[37] If it were a nuclear power plant, people would demand that it shut down!

The WHO report on Fukushima conservatively used the LNT assumption. It estimated "lifetime attributable risk (LAR)," which is "the probability of a premature incidence of a cancer attributable to radiation exposure in a representative member of the population." In the worst case—lifetime risk of thyroid cancer in female infants in the most affected areas of Fukushima province—WHO estimated a 70 percent increase in the thyroid cancer rate as a result of exposure. But this is an increased risk from a very low baseline thyroid cancer rate—below 1 percent—and so represents very few cases. "When the level of baseline incidence is that small, the actual number of 'extra' cases is likely to be small also; therefore, the impact in terms of public health would be limited." The increased cancer risk was much lower for other demographic groups and other types of cancer.[38]

In the end, then, under the most conservative assumptions and extrapolating extremely small risks to large populations, a few people *may* get cancer as a result of Fukushima, just as a few may from walking through Grand Central or flying in a jet or living in Denver. Even if true, and it depends on the no-threshold assumption, the number would be *far* fewer than the thousands who will die from the effects of burning fossil fuels when Japan shut its nuclear reactors after the accident. Nor is there any comparison to the 18,000 killed by the earthquake and tsunami. Furthermore, if our bodies actually can handle low levels of radiation, which after all vary by time and place, and low radiation is not just a small version of surviving an atomic bomb, then nobody will get additional cancer from Fukushima. In either case, the second-worst nuclear power plant accident in history was less lethal than an average coal power plant on a normal day.

Terrorist Attacks

Since the 9/11 attacks in 2001, the public imagination easily conjures up the potential dangers of terrorists crashing airplanes into things. It's easy for people to imagine that a plane crashing into a nuclear power plant would cause release of radiation if not a meltdown or a giant mushroom cloud! In reality, a nuclear power plant would be a terrible choice of target. First of all, unlike an office tower, it is low to the ground, making the task of hitting it difficult. Second, unlike an office tower, it is made not of a lattice of glass but of thick, heavily steel-reinforced concrete. By contrast, an airplane is lightweight and thin skinned. Third, the important parts—the reactor and spent-fuel pool—are often below ground level. Analysis after 9/11 concluded that a fully loaded Boeing 767 jet would do little if any damage to a nuclear reactor even if it scored a direct hit.[39]

CHAPTER EIGHT

Risks and Fears

WHY WOULD PEOPLE shut down nuclear power plants over fears of safety but allow coal plants that are far more dangerous to continue operating? Some psychological processes help explain this thinking.[1]

For one thing, people assess risk partly by how memorable or dramatic an event is. People overestimate the probability of events that are easier to imagine or more vivid.[2] Driving is far more dangerous than flying, but people fear flying more than driving because a plane crash is large scale and dramatic. After the 9/11 attacks, more people died from driving because of their fear of plane crashes than died on the hijacked planes.[3] Similarly, a nuclear power accident is dramatic, especially if everyone panics, and all the more so if a real disaster like Japan's earthquake/tsunami—or an imagined one like *The China Syndrome* film—gets cross-wired with it in our minds. (In the few years after the Fukushima accident, anxiety among Japanese citizens declined even though the risk did not change, presumably because the vividness of imagined scenarios decreased with less news coverage.)[4] By contrast, most coal deaths happen slowly and in a diffuse way, not even connected to their cause except statistically.

Not only do people generally overestimate the risk from low-probability, high-consequence events, but they do so especially if the events are dreaded—uncontrollable, unfair, and potentially catastrophic events such as plane crashes or terrorist attacks. Nuclear reactor accidents are perceived to be in this category. Fears of nuclear war, which people dread but push to the numbed-out back of their minds, also contribute.[5]

Unknown or poorly understood consequences, and delays in the effects of events, also heighten fears and exaggerate risks. Nuclear power accidents could spread

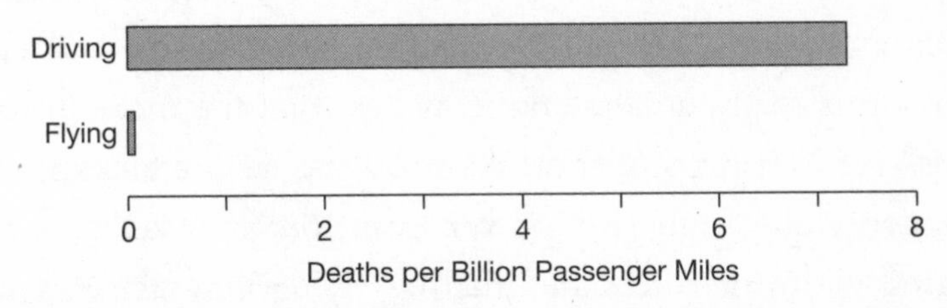

Figure 28. Deaths from driving versus flying, United States, 2015. *Source*: Ian Savage, "Comparing the Fatality Risks in United States Transportation Across Modes and over Time," *Research in Transportation Economics* 43, no. 1 (2013): 14.

invisible radiation and cause cancers years later. Merely having more information about these risks does not necessarily reduce fears, however, because anxious people may amass information selectively to buttress their preexisting fears.[6]

Nuclear power triggers risk perception on multiple dimensions. A 1987 review of the psychology of risk perception points to nuclear power as the most salient example of a disconnect between expert opinion and public perception.[7] Surveys of Americans regarding attitudes toward risk show "nuclear power to have the dubious distinction of scoring at or near the extreme on all of the characteristics associated with high risk. Its risks were seen as involuntary, delayed, unknown, uncontrollable, unfamiliar, potentially catastrophic, dreaded, and severe (certainly fatal)." By contrast, risks for things like medical X-rays—even when they involve similar amounts of radiation—were judged lower because they were more voluntary, less catastrophic, less dreaded, and more familiar.[8]

Fear of radiation also rests on people's anxiety about contamination.[9] The fact that radiation is invisible and that few of us really understand it makes it all the more insidious. Yet in truth, radiation is something we are all exposed to every day. Nuclear power contributes trivially compared with other activities such as flying in a jet, living at high altitude, or getting medical scans. Trying to eliminate all radiation because of fear of contamination is like trying to eliminate all bacteria when actually they are part of our healthy ecosystem.

The dangers of radiation form an enduring theme in popular culture, especially in movies. A large number of B-grade horror films since 1950 have used radiation as the device by which scientists gain powers that get out of control and wreak havoc. The typical effect of radiation is to make people or creatures bigger and angrier. The result has been decades of giant ants, octopuses, crabs, lizards, and blobs on the big screen, terrifying audiences and reinforcing fears of the powers of radiation.[10]

Above all, nuclear power fears are rooted in the Cold War origins of the industry and its connection to nuclear weapons. Especially for baby boomers who came of age in the early Cold War years, "fear of atomic weapons and nuclear fallout... framed the phobia about nuclear power."[11] The two actually have little to do with each other, but historically they did. Fission power was first used for a bomb and then for electric power, and the same technology that produced electric power also gave the United States material for making nuclear bombs. (The opposite is now true, as US nuclear power plants are used to safely consume nuclear weapons material produced during the Cold War.) Whatever the historical connections, a nuclear power plant cannot blow up like a bomb, and fear of nuclear weapons is not an appropriate basis for evaluating nuclear power.

Figure 29. Fears of nuclear war and radiation in the early Cold War years influenced subsequent public views of nuclear power. *Photo:* Walter Albertin / Library of Congress, 1962 via Wikimedia Commons.

Especially for the generation that grew up hiding heads under desks in fear of a nuclear attack, the very word *nuclear* is scary. If we think of the merits and risks of kärnkraft, our brains do not react the same way. Unfortunately, the blurring of nuclear power and nuclear weapons in the public mind is widespread.[12]

Regulating Risk

Consider how we regulate risk in commercial aviation. Air travel is very convenient and important to the economy, yet every once in a while, a plane crashes and kills everyone on board. When this happens, we don't all stop flying.[13] Rather, we send in investigators to figure out what happened and try to prevent it from happening again. If a part failed, the industry will inspect or replace that part in all similar planes. If pilot error occurred, airlines might implement new training procedures. As a result, flying has gotten safer and safer over recent decades.

When a company wants to build a new type of aircraft, it designs and builds the plane, gets it certified as safe, and flies it. By contrast, with nuclear power reactor designs, first the design must be approved down to the last detail, at great cost, and then built and certified, all under the careful scrutiny of government inspectors. This adds greatly to the cost. Any change of design or change of safety mandates along the way causes expensive delays and enormous piles of paperwork.

When we fly, we assume a certain level of risk. Same thing when we drive, or walk, or eat food. To reduce these risks, we have sensible government regulations and private insurance policies, and we try to behave wisely. Things are out of balance when the government tries to regulate minute amounts of materials in food that might have a slight tendency to increase disease, while in the meantime 400,000 Americans a year die from smoking. Similarly, we heavily regulate the nuclear power industry to try to make the safest source of energy in history safer still, yet we continue to allow coal, methane, and oil to change our climate and kill huge numbers of people year after year.

Nuclear power might be easier to sell to a frightened public if there were *more* accidents. Then it would look more like commercial aviation—yeah, people sometimes die, but it's way safer than the alternative. There would be relative risks to compare. But when nuclear power already is so close to zero risk, there is nothing to compare. Companies promote new designs (or governments contemplate new regulations) that can increase safety an order of magnitude—from extremely safe to very extremely safe. It's a losing proposition, because you can never reach zero risk for anything. Trying to do so just drives up costs exponentially, which is what has happened to nuclear power since Three Mile Island. In any real-world policy, zero risk means infinite cost.

Consider another example of risk, again from Sweden—the reduction of road fatalities. Sweden, the home of safety-oriented car maker Volvo, invented the modern three-point seat belt, now a global standard, and has for decades devoted enormous efforts and funds toward reducing traffic-related deaths. In the 1980s, around 1,000 Swedes died every year in traffic. In 1997 the Swedish government's "Vision Zero" plan promised to eliminate road fatalities and injuries altogether, and safety measures have now reduced deaths to around 260 per year. Safety has priority over speed or convenience, with lowered speed limits in urban areas, increased speed regulations, checks, cameras, speed bumps, improved road railings, and many other costly and sometimes cumbersome changes. These measures have succeeded, but 260 is still more than zero. "We simply do not accept *any* deaths or injuries on our roads," said the national transportation agency.[14]

At what point are massive efforts to further reduce the number toward zero simply not worth the cost, given that societies have many other social problems that deserve attention and funds? Might the efforts save more lives if directed to other proven mitigation strategies (such as reducing smoking or obesity) instead of focusing narrowly on traffic deaths?

The issue of nuclear power safety is a far more extreme example of the traffic safety problem. Already by *far* the most tightly controlled, regulated, and safest form of large-scale energy production in the world, nuclear power faces a never-ending escalation in costly safety requirements that harm both public perception and economic competitiveness. The "Vision Zero" for nuclear power has already been achieved, as illustrated by the zero death toll at Fukushima, yet *further increasing* nuclear power safety at any cost is a high priority. This is the traffic equivalent of lowering the allowable top speed for "nuclear cars" from 1 mph down to 0.1 mph, while allowing "coal cars" to drive without speed limits or brakes.

Figure 30. Coal kills. Air pollution in coal-dependent Jinan, China, 2015. *Photo*: STR/AFP/Getty Images.

In 1975 US nuclear power regulators issued a detailed risk analysis of nuclear power plant accidents, concluding that with one hundred nuclear power plants operating, the chances of any fatal accident were below 1 in 10,000, and the chances of an accident killing a thousand people were 1 in 1 million—orders of magnitude below the risks for air crashes, fires, and dam failures.[15]

Antinuclear groups strongly contested such assessments. In 1977 the Union of Concerned Scientists claimed that in fact nuclear power accidents could cause 14,400 fatal cancers by the year 2000, and the chances were 1 in 100 of an accident killing 100,000 people.[16] Although the 1979 Three Mile Island accident did not harm the public, it reinforced such fears and led to a near-shutdown of US nuclear power plant construction, with electricity being instead supplied by coal and other fossil fuels. Now, decades later, nobody has died, yet antinuclear groups continue to demand more and more reductions of already tiny risks. As a result, the world has failed spectacularly to address the real risks—coal deaths and climate change.

Indeed, the world seems to have lost a "can-do" attitude toward nuclear power that met great success in earlier decades. The earliest reactors were built quite successfully in just a few years with far fewer resources and knowledge than we now have. The US Navy put hundreds of reactors at sea with barely a problem and did it quickly, starting from scratch. Sweden and France built affordable nuclear power plants that today's efforts seem unable to match. We now use computers instead of slide rules, we have much better-educated populations, our GDPs have multiplied, and we have engineering methods and materials that our grandparents could only dream of. Despite all that, much of the world seems to be frozen in fear—fear that is not reality based but that is really killing us.

"The ultimate question," wrote physicist Alvin Weinberg after Chernobyl, "is whether we can accept a technology whose safety is measured probabilistically. Members of the nuclear community had always assumed that if the probability of a severe accident was sufficiently low, nuclear power would be accepted."[17] This assumption of the public's ability to assess risk was perhaps unrealistic.

Of the many people who feel that nuclear power is just "too dangerous," surprisingly few ever ask, "Compared to what?" It's pretty dangerous compared with fairy dust, which meets the world's growing energy needs without any costs or risks. But compared with the world's leading, and fastest-growing, power source—coal? Compared with oil and gas?

In thinking about nuclear power safety, one should always ask, "Compared to what?" And the answer is: compared to coal—the world's dominant and fastest-growing fuel, the leading cause of climate change, the fuel that kills a million people a year. Compared to *that*.

Scary Versus Dangerous

Scary and dangerous are not the same thing.

Jumping off the top Olympic high-dive platform, 33 feet high, would be scary for many of us. But it would not be particularly dangerous, assuming you can swim. The risks—of landing badly and injuring ourselves or even panicking and drowning—are not zero but quite small.

But now imagine you are on a long railroad bridge, 33 feet above a deep body of calm water, and a train is coming across the bridge at you. The jump is still scary. But it's the train that's actually dangerous. If you freeze in fear and don't jump, you will die. And if you start running away from the train, figuring that although you can't make it off the bridge, at least you are moving in the right direction, you will also die.

This is humanity's situation: Climate change is the train coming down the track at us, likely to cause catastrophic harm. The popular responses—ramping up renewables, moving from coal to methane, and such—move us in the right direction but do not get us off the bridge in time. We have a solution that works, that Sweden and others have proven, and that's no more dangerous than jumping off that bridge, but it's scary. That solution is to rapidly expand our use of nuclear power.

Figure 31. Jumping is scary; staying on is dangerous. *Photo*: Drew Jacksich via Wikimedia Commons (CC Attribute-Share-Alike Generic 2.0).

For many people, nuclear power is scary, but it has an extraordinary safety record over fifty years. Understanding the difference between scary and dangerous, between jumping in the water and being run over by a train, may be the key to our future.

CHAPTER NINE

Handling Waste

IN NO ASPECT of nuclear power do fears and realities diverge more than the question of radioactive waste. The quantities are so much smaller than the waste generated by any other fuel-using power source that they are easily managed. For the electricity used by an American in an entire lifetime, if generated entirely by coal, the solid waste would weigh 136,000 pounds. If generated entirely by nuclear power, that lifetime of electricity would produce waste weighing 2 pounds that would fit in a soda can, only a trace of which would be long-lived waste.[1]

Most radiation in nuclear waste decays quickly and becomes safer over the years. As we will see, current interim storage methods are safe, relatively inexpensive, and practical for the next century. But some countries, including

Sweden, have sought to allay fears by planning elaborate long-term storage systems. Sweden's system is operated by a company, the Swedish Nuclear Fuel and Waste Management Company (SKB), owned and funded entirely by the nuclear power companies themselves.[2] (Like Sweden's nuclear power stations, all the SKB facilities are open to the public and give tours to thousands of visitors every year.) After spent nuclear fuel has cooled for a year in water pools at the power plants, becoming less radioactive, SKB transports it in its own ship to an interim storage facility near one of the power stations. The radioactive spent fuel is stored there in pools of water 26 feet deep—the water absorbs the radiation safely—located in bedrock 100 feet below ground level. At this facility are stored about 7,000 tons of spent

Figure 32. Sweden's interim spent-fuel storage. *Photo*: Courtesy of Curt-Robert Lindqvist / SKB.

fuel, which is the entire high-level waste from Sweden's nuclear electricity generation for more than 40 years.

SKB's design for final disposal of spent fuel uses a planned encapsulation plant next to the interim-storage facility.[3] After the waste has cooled down and become less radioactive during its stay in the interim-storage pools, it will be inserted into cast iron that will be encapsulated in copper canisters sealed with a specialized welding method. These 25-ton canisters will be transported by SKB's ship about 250 miles to the final repository, where they will rest in tunnels in bedrock 1,600 feet underground, packed in clay. The rock has been stable for almost 2 billion years and is expected to hold the spent fuel safely away from groundwater for 100,000 years until the radioactivity has decayed away, even if conditions change such as with sea-level rise or an ice age. Over 50 years, SKB expects to store 6,000 canisters at the site. SKB hopes to begin construction of the final repository in the early 2020s and complete it in about ten years.

For shorter-lived low- and intermediate-level nuclear waste, from both power plants and medical and industrial facilities, Sweden has for decades operated a final repository near the planned high-level waste repository. To keep the waste isolated for 500 years, the facility stores it in rock vaults and a concrete silo, 150 feet underground.

Although Sweden has not begun construction on its final repository, next-door Finland has adopted Sweden's method and is the first country in the world to begin construction on its repository for long-lived nuclear waste.[4] The facility is close to one of Finland's nuclear power stations. It is scheduled to begin storing spent fuel in 2022 and is expected to operate for 100 years until full. Safety studies have estimated the effects if the repository were to fail and leak. In the most pessimistic scenario, the copper canister is already damaged before being buried and the clay disappears in a thousand years (both extremely unlikely), and a person then lives his or her entire life on the most contaminated square meter of soil, eating and drinking only food and water from that same square meter. In that scenario the person would receive annual radiation equal to eating a bunch of bananas (.00018 mSv/year).[5]

The Finland failure scenario illustrates a key point. Many people worry about long-term storage of nuclear waste as though any amount of radiation, any contamination of the environment, would be catastrophic. As we have seen, the environment is already filled with radiation, and small amounts of contamination over time would not be a big problem. Zero risk is not necessary. Furthermore, as years pass, the radiation levels decrease, unlike toxic elements in waste from coal (or from batteries and solar panels for that matter), which last essentially forever and will surely spread into the environment over time. The real danger is that we freak out at the possibility of a tiny leak of radiation many years in the future, shut down the nuclear power industry because we cannot preclude that possibility at reasonable cost, and then go on burning fossil fuels.

In the urge to quarantine nuclear waste forever, few people consider how we treat other toxic wastes. Many industrial activities produce large quantities of highly toxic and lethal waste that cannot be treated and do not have "half-lives"—materials that must be handled and stored safely *forever*. In general, the requirements, if any, for safe storage of *these* materials are far more relaxed than for the comparatively harmless spent nuclear material. The Swedish mining company Boliden is currently constructing a "world unique" repository for 400,000 tons of extremely toxic "forever" waste that includes large quantities of mercury and arsenic.[6] The facility is "unique" since, according to Boliden, no other similar industry in the world has any plans of long-term safe storage of this kind of waste.[7] The cost of the 400,000-ton "forever repository" is estimated at about $50 million, or about $120/ton of stored waste. In comparison, the time-limited nuclear materials storage facility for 12,000 tons of fuel that is proposed by SKB in Sweden is expected to cost at least $17 billion, or about $1.5 million/ton of waste stored.[8] Thus, the requirements for time-limited storage of nuclear waste cost about ten thousand times more per ton than forever storage of far more dangerous chemicals (in the rare case where this is done at all).

Yet other countries continue to plan extremely expensive facilities for extremely long-term nuclear-waste storage.

Like Finland, France is testing a similar, and larger, underground permanent storage facility, although designs are not finalized.[9]

Canada's high-level nuclear waste in total, accumulated since the 1960s, if stacked together, would fit in seven hockey rinks, each stacked less than four feet high.[10] The spent fuel rods are stored in water for seven to ten years, becoming less radioactive, and then transferred to dry storage in casks with concrete containing the radiation. Canada has begun looking for a location to build a long-term underground repository.

In the United States too, nuclear waste is readily managed because the volumes are so small. The entire volume of spent fuel from fifty years of American nuclear power—a source that produces one-fifth of US electricity—could be packed into a football stadium, piled twenty feet high.[11]

Of course, some people do not want that particular stadium in their backyard. The United States mismanaged the policy issue of nuclear waste disposal. It mandated a single, extremely costly, disposal site, Yucca Mountain, located in Nevada. Nuclear waste would be shipped to Nevada from around the country. Yucca Mountain was supposed to hold the waste safely, with no risk of leaching into the surrounding ecosystem such as groundwater, for tens of thousands of years. That is almost impossible to guarantee, even with layers of protection in deep underground caverns surrounded by natural layers of salt. The standard of zero risk, again, makes

Figure 33. Yucca Mountain tunnel, 2014. *Photo*: Nuclear Regulatory Commission (CC BY 2.0).

success essentially impossible. After the US senator from Nevada became the majority leader of the Senate, the government killed off the facility after investing more than $15 billion in the project. In 2017, after a change of administration, efforts to restart the Yucca Mountain repository began.[12]

The US military also has nuclear waste to dispose of, from its nuclear weapons and ship-propulsion programs. With little of the drama of Yucca Mountain, it built a deep underground repository in New Mexico and operated it successfully for decades with just one serious accident in 2014 that triggered an expensive cleanup but did not harm human health or the environment.[13]

One way to isolate waste safely for hundreds of thousands of years would be to bury it under the seabed, miles below the water's surface, in areas that are almost completely lifeless and inert.[14] This approach was proposed decades ago but ultimately rejected after an outcry by environmentalists. (Against the hypothetical chance of an accident, in an ocean that contains natural uranium anyway, we have the reality of ongoing oil spills at sea since the world relies on oil instead of nuclear power.)

Given all the drama and extreme expense of long-term waste disposal, in truth there is no imperative to put nuclear waste into permanent storage anytime soon. In a hundred years, most of the radioactive elements around today will have decayed away, and the volumes of very long-lived radioactive elements will be small. The thing we fear in nuclear waste—radioactivity—is what helps solve the problem. Radioactive elements have a half-life, during which time the radioactivity decreases by half. Wait that long again, and it's cut in half again. Most nuclear waste decays away to very low radioactivity as it sits for a year or two in a cooling pool after being removed from a reactor.

This is not true of, say, coal waste—both solid and airborne—which remains as toxic centuries in the future as it was when first dumped into the water or air. By the way, coal waste contains radioactive elements. You will get a higher dose of radiation living next to a coal plant than a nuclear power plant. Even the waste from solar cells, when their useful life is over in twenty-five years, remains toxic for many decades and requires special handling for disposal or reuse.

Not all nuclear waste has a short half-life. A residual part, after the short-lived elements have decayed away, will remain radioactive for very long times, tens of thousands of years for some. These are the elements that capture the public imagination and induce fear. But suppose we just put that problem off for a century, leaving spent fuel safely stored in concrete casks and letting future generations solve the problem more permanently. Although this might seem irresponsible, it is nowhere near as irresponsible as leaving the next generation an overheated planet out of control. And actually it would be reasonable because technology in a hundred years will be better able to solve the problem.

In fact, people today are already developing nuclear power designs that use this high-level waste as fuel. The reason nuclear waste has such long-lived radioactivity is that today's power plants require removal of the fuel while the vast majority of its energy is still unused. Some of the new fourth-generation designs discussed in Chapter 12 could tap this energy while using up the waste we have stored from today's reactors. Breeder reactors that can make use of spent fuel as new fuel are already in operation today in Russia and elsewhere.

What the United States actually is doing, currently, is to just store nuclear waste on-site, at operating and closed

Figure 34. Dry casks containing high-level radioactive waste. *Photo*: Nuclear Regulatory Commission.

nuclear power plants, in large concrete "dry casks." Spent fuel is left for a couple of years in cooling pools to let the shorter-lived elements decay and then transferred to the casks and placed on a concrete pad to sit and wait for politics to catch up with them. According to the Nuclear Regulatory Commission (NRC), no radiation leaks have occurred in the three decades of dry-cask use so far, and the casks provide safe storage for at least 120 years.[15]

A visit to the recently closed Vermont Yankee nuclear power plant underlines the point about waste. A visitor can stand next to the 19-foot-high, 12-foot-diameter cylindrical concrete casks, sitting on a simple flat pad high above the flood level of the Connecticut River that cooled the plant. There are several dozen casks, and as the remaining fuel is transferred, the number will grow to fifty-eight. Together they take up a surprisingly small space, and you can stand near them with no radiation protection because most radiation does not pass through concrete. The setup is surrounded by layers of armed security, far more than warranted by any actual threat. The nuclear waste is stable, secure, and incredibly compact, considering that it is the entire high-level radioactive waste output from the 42 years that the 650 MW plant operated.

From a political and public-opinion perspective, storing nuclear waste is an issue. But from a scientific and technical perspective, it is a nonissue. Spent nuclear fuel has now been stored across the globe for nearly 70 years with barely an incident,[16] an enviable track record for any type of industry. You can build a long-term repository, as Finland is doing, or leave it sitting out in the backyards of nuclear power plants in concrete casks, as the United States is doing. It is not an urgent problem the way climate change is an urgent problem.

CHAPTER TEN

Preventing Proliferation

NUCLEAR FISSION, IN which a tiny amount of mass is converted to a large amount of energy by splitting a big atom, is the basic power source of both nuclear power plants and nuclear weapons. But they are not the same thing, any more than using gasoline in your car is the same thing as dropping napalm, though both use the same basic chemistry.

If you hear the word *nuke* or see a protest sign reading "No Nukes," do you think of nuclear weapons or nuclear power plants? It could be either, and that's a problem. Nuclear weapons are extremely scary and dangerous, have killed hundreds of thousands of people historically, threaten billions more today, and do not have a useful purpose except possibly in a negative sense of deterring their use by others. Nuclear power, by contrast, has been extremely safe and has saved millions of lives through the displacement of coal and other deadly fossil fuels.

There are, nonetheless, connections between the two "nukes," and we will need to take these into account if we are to successfully use nuclear power to help rapidly decarbonize the world.

One connection is historical. The United States and the Soviet Union mastered fission during World War II and the Cold War with the primary purpose of building a lot of big bombs. The nuclear power industry produced electricity, but in some places it also produced by-products that served as fissile material for nuclear weapons. The very earliest reactor designs were chosen to maximize the production of these elements—exactly the opposite of what we want now from reactor designs.

Uranium occurs naturally around the world, but in its natural form it contains only a little bit of the isotope[1] needed for fission. Through great effort, with thousands of centrifuges feeding from one to another, it is possible to concentrate the potent variant of uranium enough to build a bomb from it. This is called uranium enrichment. This is what Iran was doing, for example, before the international deal that froze its enrichment program so that it couldn't build nuclear weapons. Pakistan has also used centrifuges to build uranium bombs, and North Korea has done the same.

Figure 35. Uranium-enrichment centrifuges seized in 2003 en route to Libya, which had no civilian nuclear power program. *Photo*: US government, Y-12.

A second material suitable for building a bomb is certain isotopes of plutonium. Plutonium does not occur naturally but can be produced in a nuclear reactor as a by-product of splitting uranium atoms. This was the primary route to nuclear weapons for the great powers, including the United States. During World War II, the United States created both a uranium bomb and a plutonium bomb and dropped them on Hiroshima and Nagasaki, respectively.

The key task for preventing the spread of nuclear weapons is to keep these two materials—highly enriched uranium (HEU) and plutonium—away from new countries or nonstate actors. Obtaining plutonium is not too difficult if one has certain kinds of nuclear reactors. For example, Japan has a pile of it right now. But building a plutonium bomb is quite difficult. A terrorist couldn't do it, and even a sizable country such as North Korea needs a concerted program over many years and using large budgets. By contrast, building a uranium bomb is much simpler, but obtaining HEU is difficult.

When nuclear power first emerged as a useful source of electricity around the world, the international community put in place strong governance structures to ensure the peaceful uses of nuclear materials. Above all, the countries of the world wanted to prevent a country from secretly diverting plutonium or HEU from a nuclear reactor into a weapons program. To this end, the world established the International Atomic Energy Agency (IAEA), an independent arm of the United Nations, with the mission of inspecting nuclear power programs around the world and making sure materials were handled safely and peacefully.

The IAEA has been remarkably effective. Its inspectors have the power to make intrusive inspections, to leave cameras in place watching what happens, to place seals on containers so that they can't be secretly opened, and to interview scientists in order to make sure things are what they appear to be. At certain important locations, the IAEA has its own full-time staff and its own independent labs inside of nuclear facilities to ensure that complete and uninterrupted safeguards are in place.

It's hard to keep nuclear secrets in today's world. When Iraq's Saddam Hussein was suspected of developing a nuclear weapons program, IAEA inspectors crawled all over the place and didn't find anything only because Saddam had shut the program down years earlier. When Iran enriched uranium, the world found out about it. When North Korea made a deal to shut down its plutonium production but kept running a secret uranium-enrichment program, the world found out about it. (North Korea then withdrew from the IAEA in order to carry on.) When a Pakistani scientist sold uranium-enrichment technology to several countries, the

Figure 36. IAEA inspectors installing monitoring equipment, Czech Republic, 2015. *Photo*: International Atomic Energy Agency.

world caught him at it. When Syria built a secret nuclear reactor, Israel bombed it.

Decades ago, it was widely feared and assumed that by now dozens of countries would have nuclear weapons, but this has not happened. Nine countries have them, including the problem case of North Korea, but proliferation has not gone beyond that.

The main reason more countries do not have nuclear weapons is that they have chosen not to. Dozens of countries have the ability, but they have concluded that the costs far exceed the benefits. Sweden once considered nuclear weapons but decided not to pursue them. Argentina and Brazil started in on an arms race, but stopped before either built its first weapon. South Africa actually built a nuclear weapon but then destroyed it and scrapped the program when apartheid ended.

These countries, and almost all others, belong to a treaty, the Non-Proliferation Treaty (NPT), that requires them to forgo nuclear weapons. The NPT/IAEA system has successfully decoupled nuclear power technology from the spread of weapons. The only nonmembers of the NPT (putting aside South Sudan, which has been in a civil war since its recent founding and has no nuclear interests) are India, Pakistan, and Israel—the three states that developed nuclear weapons after the original five were grandfathered into the treaty. In addition, North Korea belonged to the NPT but then withdrew from it, the only nation ever to do so, and built nuclear weapons.

In the past thirty years, the superpower arms race has gone into reverse, with the stockpiles of tens of thousands of nuclear weapons reduced on both sides by 75–80 percent. The scientific and engineering complexes that had been devoted to building weapons have now mostly shifted to dismantling them. This means that the civilian nuclear power industry now has less and less connection to the military, which no longer needs more plutonium. In fact, a lot of enriched uranium from dismantled Soviet warheads was watered down and used as fuel in civilian US nuclear reactors. From 1998 to 2013, about 10 percent of the electricity used in the United States came from 20,000 dismantled Soviet warheads.[2] Additional uranium and plutonium from weapons has been converted for use as fuel in American and Russian civilian reactors and naval propulsion.[3]

Nuclear power has not been a factor in the proliferation of nuclear weapons to new countries.[4] The countries that have refused to join the NPT, have kicked out IAEA inspectors, or have secretly built their own nuclear weapons have not done so using civilian nuclear programs. Neither Israel nor North Korea even has a civilian nuclear power program. As five experts on proliferation put it in 2000, "To date, commercial nuclear power has played little, if any, role as a bridge to national entry into the nuclear arms race, nor are there any known cases in which individuals or subnational groups have stolen materials from nuclear power facilities for use in weapons."[5]

Actually, the creation of civilian nuclear power programs in a number of countries has served as a check on nuclear weapons proliferation. Countries with nuclear know-how have helped new countries to build and operate nuclear reactors for electricity, and those new countries in turn have signed the NPT and accepted the strict rules of the IAEA.[6] This happened in South Korea, for example, in the 1970s, when it decided against having a bomb and instead turned to developing a civilian nuclear power industry under international supervision. Today, it is a model of affordable and safe nuclear power—even exporting power plants to the United Arab Emirates (UAE) recently—and still does not have a nuclear bomb despite the bad behavior of its threatening neighbor, North Korea.[7]

If the number of nuclear reactors in the world increases, and more countries host them, the international community can and should expand its efforts to ensure that reactor fuel does not get processed into potential bomb material.[8] A key method to do this is to keep countries from building their own uranium-enrichment and fuel-reprocessing facilities. For reactor use, uranium has to be enriched from below 1 percent of the fissile type to 4–5 percent (known as low-enriched uranium, or LEU). LEU cannot be used for bombs. (The 2015 multinational agreement with Iran limits its uranium enrichment to 3.7 percent.)[9] But countries that master enrichment methods themselves might be tempted to enrich uranium to above 90 percent (HEU) for weapons use. Countries that invest large sums in long-lived reactors want assurance of a supply of LEU, the fuel for the light-water reactors that predominate today. Hence, the international nuclear power regime provides LEU externally to countries that own reactors but not nuclear weapons.

The IAEA has recently created a physical LEU fuel bank that will assure countries of access to fuel in the event of unusual circumstances in which their regular supply is disrupted, but only when comprehensive IAEA safeguards are in place for that country. The bank, located in Kazakhstan and set to receive LEU in 2018, will have enough fuel to power a city for several years. It may never be needed but is

Figure 37. IAEA Kazakhstan LEU fuel bank, 2017. *Photo*: Courtesy of Nuclear Threat Initiative.

intended to prove that countries with reactors do not need their own nuclear fuel-cycle infrastructure with its potential to be diverted for misuse.[10] (Russia also operates its own physical LEU bank, and the United Kingdom offers a guarantee of LEU supply.) Kazakhstan, incidentally, is a model country that inherited about 1,500 nuclear weapons when the Cold War ended but got rid of all of them within a few years. It is also the world's largest miner of uranium.

Further initiatives have been proposed to ensure nonproliferation in a world of growing nuclear power. In several proposed ways, the international community could internationalize the fuel cycle. For example, Daniel Poneman, an expert on nuclear fuel and proliferation, has proposed an Assured Nuclear Fuel Services Initiative that would guarantee reactor fuel at good prices to countries that commit to not seek enrichment or reprocessing capabilities. He argues that this approach would be more effective than current unilateral US efforts to impose stringent safeguards, which only lead buying countries to find fuel from less demanding suppliers.[11]

Incidentally, a source of confusion around nuclear power and weapons is the use of nuclear power for propulsion in some warships. These nuclear-powered ships may carry nuclear weapons such as missiles or may carry no nuclear weapons, and a non-nuclear-propelled ship may similarly carry nuclear weapons or not. There is no connection between the type of propulsion and the weapons on the ship.

Naval nuclear propulsion was a clever accomplishment of Admiral Hyman Rickover in the early Cold War. He built nuclear reactors quickly and successfully to power submarines and aircraft carriers in particular. This concentrated power source allows these warships to remain at sea, including underwater, for long periods. Worldwide, nuclear-powered ships have logged more than 12,000 reactor-years of operation with 700 reactors, 200 of which are currently operating, and the last serious reactor accident was more than thirty years ago. (As with civilian reactors, the only lethal naval reactor accidents were in the Soviet Union, which had several in the period 1961–1985. The US Navy has operated 6,200 reactor-years with no known radiological incident.)[12]

A final consideration making nuclear proliferation easier to prevent is that countries are not fighting each other currently. Although the world's countries are armed to the teeth, many conflicts simmer, and a war between countries could break out anytime—in 2018 the US–North Korea standoff was especially scary—such wars have become quite rare. All today's active wars are civil wars, within countries. Historically, regular national armies fought against each other constantly. During the Cold War, the most destructive and deadly wars were between those national armies—the Korean War, India-Pakistan, and Iran-Iraq, for example. But today, the last such war was the invasion of Iraq in 2003,

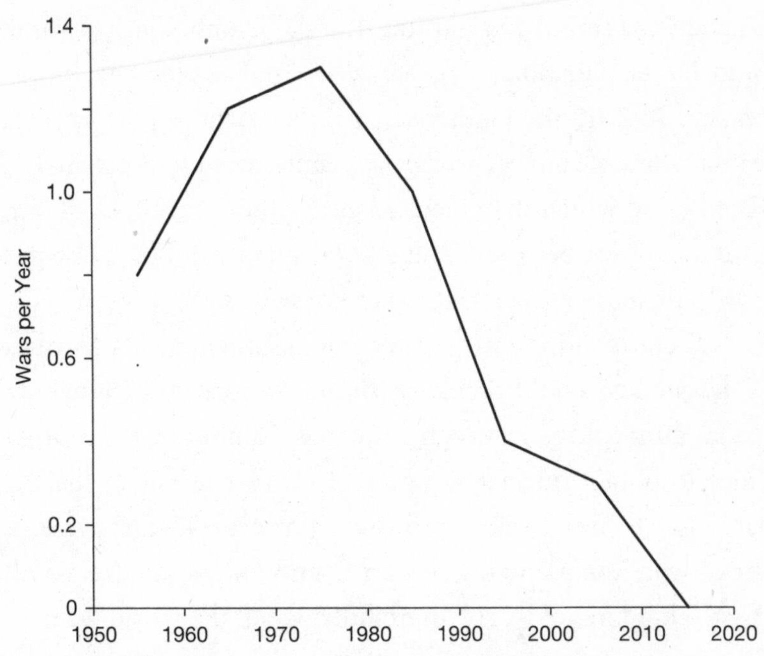

Figure 38. Interstate wars by decade. *Data source*: UCDP/PRIO Armed Conflict Dataset, version 17.2. Wars defined as armed conflicts with more than one thousand battle fatalities in the year. See Marie Allansson, Erik Melander, and Lotta Themnér, "Organized Violence, 1989–2016," *Journal of Peace Research* 54, no. 4 (2017); and Nils Petter Gleditsch et al., "Armed Conflict, 1946–2001: A New Dataset," *Journal of Peace Research* 39, no. 5 (2002).

a decade and a half ago.[13] Whole categories of warfare such as tank battles and major naval battles are also disappearing.

These days, when two hostile countries skirmish, the violence tends to die down quickly rather than escalate to all-out war. Some frightening episodes have brought state armies into small-scale violent clashes, such as in

Armenia-Azerbaijan, Ukraine-Russia, Cambodia-Thailand, and Israel-Lebanon in recent years. But cooler heads prevailed. In 2017 the Chinese and Indian armies performed a ritual standoff in a nearly inaccessible mountaintop area of Bhutan, in which they insulted each other and threw sticks. But a real war between China and India would be a disaster for both, and everyone seems to know this.

Of course, this state of nonwar between regular armies is fragile and could break down, but it does still contribute to an atmosphere in which countries do not feel great pressure to obtain nuclear weapons. Contrast this with the US decision to build the bomb during World War II when a huge war was going on and an enemy might get the bomb first. That kind of world, mercifully, is not the world we now live in.

As we will discuss in Chapter 12, new nuclear power designs in the coming years will make it even harder to divert material for military use. Meanwhile, we can be reassured by an impressive record over the decades since nuclear technology was developed. As with accident safety, so with proliferation: there is always plenty to worry about, but in practice the system works. With continuing effort, we can keep risks extremely low.

THE WAY FORWARD

Humanity can fix climate change *and* prosper by expanding on the example of countries like Sweden and France: building out both nuclear power and renewable energy, developing new reactor designs, and pricing carbon pollution.

CHAPTER ELEVEN

Keep What We've Got

IF THE UNITED States had built out the fleet of nuclear power plants it once planned, we would be much closer to solving climate change today. Largely because of the agitation of antinuclear groups such as Greenpeace, most of the planned nuclear power plants were never built, and quite a few that had been partly built were abandoned. A wave of new planned nuclear power plants in the early 2000s[1] has also been scrapped since the 2011 Fukushima accident.

Recently, these groups have expanded their antinuclear goals and are successfully lobbying and lawyering to shut down existing nuclear power plants before the end of their useful lives. We have seen how Germany is doing this in a post-Fukushima panic and how that decision has stalled

Germany's progress in reducing CO_2 emissions (see Chapter 3). The story is the same elsewhere.

With cheap fossil fuels available as an alternative—methane or coal, depending on the country—power producers can hardly be blamed for bailing out on nuclear power after facing nonstop litigation, regulation, and agitation from antinuclear groups. But in every case, nuclear power capacity has been at least partially replaced by fossil fuels, and CO_2 emissions have risen as a result. Back in the 1970s, people did not understand climate change, and had far less experience with the safety of nuclear power, so antinuclear activism was more understandable. Today, the context has completely changed, but the antinuclear groups have not updated their perspective.

One of the least expensive clean-energy sources in the world is an existing nuclear power plant already generating electricity today. Nuclear power plants are expensive to build but inexpensive to operate, so once you have one working fine, it makes sense to keep using it. There are 449 power reactors operating around the world, 99 of them in the United States, where nuclear power supplies 20 percent of electricity.[2] In the United States, that is more than hydropower, wind, and solar combined. Worldwide, nuclear power is second only to hydropower among carbon-free energy sources.[3] A no-brainer way to combat climate change today is to start by simply *not* closing down existing nuclear power plants. Shutting down a leading low-carbon electricity source at this critical moment in history is a step backward in solving the climate problem.

Pressure to close existing nuclear power plants does not come from frightened residents near the plant. On the contrary, polling shows that people living closest to a plant are most supportive, with 83 percent support among those living within ten miles of a US plant and 82 percent among those living within 30–50 miles in Sweden.[4] (Also, more than three-quarters of Americans who consider themselves well informed about nuclear power support it, versus about half for those who consider themselves not well informed.)[5] Those who live closest understand a nuclear power plant best, benefit most from its jobs, and see they have little to fear. Those who live farther away do not understand the science and engineering, or the safety plans, and their fears fill that void.[6] These fears among the uninformed public are the raw material with which well-funded antinuclear organizations fuel their campaigns to close existing plants.

Vermont Yankee

Consider Vermont and western Massachusetts, where, until recently, most of the carbon-free electricity was supplied by the Vermont Yankee power plant, located near the Massachusetts border. The region had been a hotbed of antinuclear activism ever since the movement began in the 1970s as a campaign to stop two nuclear reactors planned for a site in Massachusetts just downriver. (The utility unwisely

Figure 39. Vermont Yankee Nuclear Power Plant. *Photo*: Nuclear Regulatory Commission.

sited the planned nuclear power plant near three hippie communes, sparking opposition that spread to neighboring towns and eventually led to the plant's cancellation.)[7]

In 2010 the Vermont Legislature passed a law against Vermont Yankee's continued operation, and that same year a new governor was elected who had led the opposition to the plant. Despite the relicensing of the plant for twenty more years in 2011, the owner announced in 2013 that Vermont Yankee would close the next year.

Vermont Yankee had been providing electricity wholesale at about 4 cents/kWh and tried unsuccessfully to negotiate a new Power Purchase Agreement (PPA) with Vermont at 4.5 cents.[8] This is not quite as cheap as methane at current low prices (since fracking). But it is far cheaper than clean alternatives such as wind and sun. In fact, at the same time, Massachusetts was negotiating an agreement for offshore wind power at almost 19 cents/kWh.[9]

Vermont Yankee faced competition from subsidized energy sources. Fossil fuels are subsidized and not just by being allowed to dump CO_2 in the air with no charge. An estimate of US government energy subsidies up through 2003, about $650 billion, breaks down as 47 percent to oil, 13 percent each to coal and gas, 11 percent to hydropower, 10 percent to nuclear power, and 6 percent to renewables.[10]

In recent years, renewables have been strongly subsidized by both federal and state governments. Many states mandate that utilities include a certain amount of renewables—not a certain amount of carbon-free generation—and nuclear power is not included. For example, in Vermont in 2013, in addition to the state's share of $15 billion in federal subsidies for renewables, state-level subsidies included customer credits, grants, loans, feed-in tariffs (price guarantees), investment-tax credits, property-tax abatements, sales-tax exemptions, and buybacks of excess capacity at peak times.[11]

In the context of subsidies for both fossil fuels and renewables, a plant such as Vermont Yankee has a hard road. Methane is cheap, and renewables are cool. Every move gets scrutinized by regulators and protested by activists. It doesn't help that the companies that own nuclear power plants often also own fossil plants. (The owner of Vermont Yankee operates one-third nuclear power plants and two-thirds fossil fuel.) If they replace a nuclear power plant with a methane plant, they can compete better and sell more electricity. The Vermont state government would not sign a PPA to guarantee a steady market and price for nuclear-generated electricity. At the end of 2014, Vermont Yankee closed after forty-two years of operation, with seventeen years left on its operating license.

In western Massachusetts, when Vermont Yankee went offline, electricity prices spiked. As methane filled in to produce electricity, the gas company demanded to build new methane pipelines to supply the region. Climate activists argued, correctly, that investments in that kind of fossil-fuel infrastructure, which would last for decades, would lock us into a fossil economy far into the future. They marched against the gas pipelines and had some success in delaying and canceling them. So the gas company imposed a moratorium on new gas hookups for communities along the Connecticut River. Business suffered, and new construction had to use the more expensive fossil-fuel propane instead of methane. Two years later, that's where things still stood.

Transmission lines to bring in Canadian hydropower were another planned way to offset the loss of Vermont Yankee (and the Massachusetts coal plants that shut down). But the largest-scale attempt to add renewable electricity for Massachusetts, 1100 MW of hydropower from Canada, was rejected in 2018. After a seven-year, $250 million application process, New Hampshire regulators saw the power lines through their state as an eyesore.[12]

During a prolonged cold spell in December–January 2017–2018, when New England temperatures remained below freezing for weeks at a time, the grid suffered from the absence of Vermont Yankee. With methane supplies diverted to home heating needs, the grid switched to oil as the leading fuel for electric generation. Supplies of fuel oil, kept at generating stations for such a contingency, dwindled to alarmingly low levels over several weeks. Oil trucking capabilities were stretched to the maximum, and icebreakers opened routes for oil deliveries by sea. Carbon emissions rose. LNG stores supplemented pipeline gas, but supplies remained extremely tight. Natural gas prices in Massachusetts peaked briefly at *twenty times* the price just a month earlier. The price of electricity jumped fivefold.[13] Solar power was diminished seasonally (winter) and dropped to near zero at times, owing to snow. The closing of Vermont Yankee was not the main factor in this situation but contributed to it.

Massachusetts now plans to close its last remaining nuclear power plant, Pilgrim, in 2019, with thirteen years still left on its license. Much more electricity generation will be lost with the closing of this one nuclear power plant than Massachusetts generates with all its solar, wind, and hydropower combined.[14]

After Vermont Yankee closed, CO_2 emission rates rose across New England, reversing a decade of declines.[15] Emissions for all of New England rose almost 3 percent within a year. This rise amounted to a million more tons of CO_2 per year added to the atmosphere. The problem is not so much the increase but the failure to decrease. If a politically liberal, rich, and technologically advanced region such as New England cannot rapidly decarbonize, what hope is there for the world as a whole to do so?

New England still gets 30 percent of its electricity from nuclear power, but with the last plant in Massachusetts set to close in 2019, only one in Connecticut and one in New Hampshire will remain. As the last Massachusetts plant goes offline, it will be replaced with methane and oil, and emissions will again rise, or at least fail to decline.

Other US States

California, like Massachusetts, is closing all its nuclear power plants. It closed the 2 GW San Onofre plant in 2013 and missed its CO_2 emissions targets as a result. Methane, again, replaced the lost capacity.[16]

Recently, the state announced a big agreement among the utility, the antinuclear groups, the labor unions, and the state to close the last nuclear power plant, Diablo Canyon. The plant generates 9 percent of California's electricity and has operated successfully for thirty-one years. The

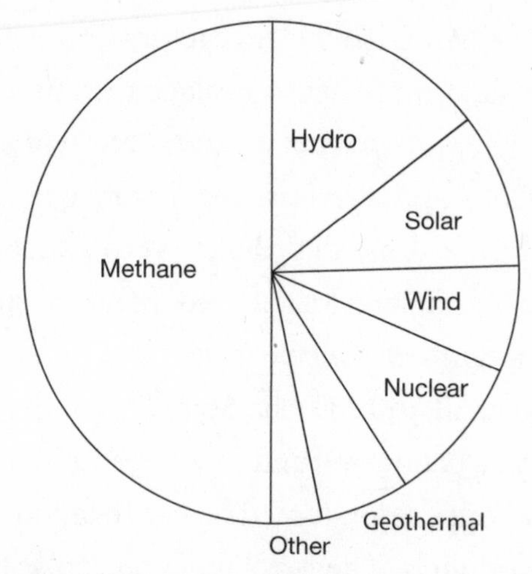

Figure 40. California energy mix, 2016, after the San Onofre nuclear power plant closed but with Diablo Canyon still operating. *Data source*: California Energy Commission.

agreement declares that Diablo Canyon will be replaced with clean renewable power, but this is misleading.

First of all, any renewable power that is built could be replacing fossil fuels if it were not replacing nuclear power. This is the same situation we saw in Germany.

Second, despite its lofty declarations to replace Diablo with clean energy, the agreement does not specify how this will happen. It states that "the Parties cannot, and it would be a mistake to try to, specify all the necessary replacement procurement now."[17] The day that Diablo Canyon closes down, but Californians do not stop using electricity, it appears likely that methane will fill the gap.

The most recent US nuclear reactors under construction, in South Carolina, were set to replace the current coal generation supplying electricity to that state. However, under the pressure of cost overruns, regulatory gum-ups, delays, and political pressures, the plants were canceled in 2017, with billions of dollars lost. (Two reactors at a plant in Georgia remain under construction.)

On Long Island, in 1989, the $6 billion Shoreham nuclear power plant was completed and set to open but instead closed when political opposition forced its cancellation and replacement with fossil fuels. The fossil fuel that replaced Shoreham has, in the intervening years, contributed on the order of 80 million tons of CO_2 to the atmosphere.[18] That's the weight of about 40 million Ford Explorers, also known as a lot of CO_2.

Several US states have recently decided to give subsidies to existing nuclear power plants similar to the credits given to renewables. Illinois, New York, and New Jersey have saved large nuclear power plants in this way, and Connecticut is moving in the same direction. New York's action, however, is a mixed bag, as it saves several reactors upstate but shuts down the 2 GW twin reactors at Indian Point that have powered New York City for decades.[19]

Nationally, a capacity of 5 GW, more than the Swedish power station at Ringhals (see Chapter 1), was lost to nuclear power plant retirements in 2013–2017. The US government expects a similar decrease in the next nine years.[20]

Other Countries

Alarmingly, the trend to shut down existing perfectly good nuclear power plants seems to be spreading around the world. Since the Fukushima accident in 2011, politicians in many countries have concluded that opposition to nuclear power and fears of disaster are forces they do not wish to stand up against. Sadly, there is almost no public opposition or political price to pay for approving a new methane gas power plant or, in many countries, a new coal plant.

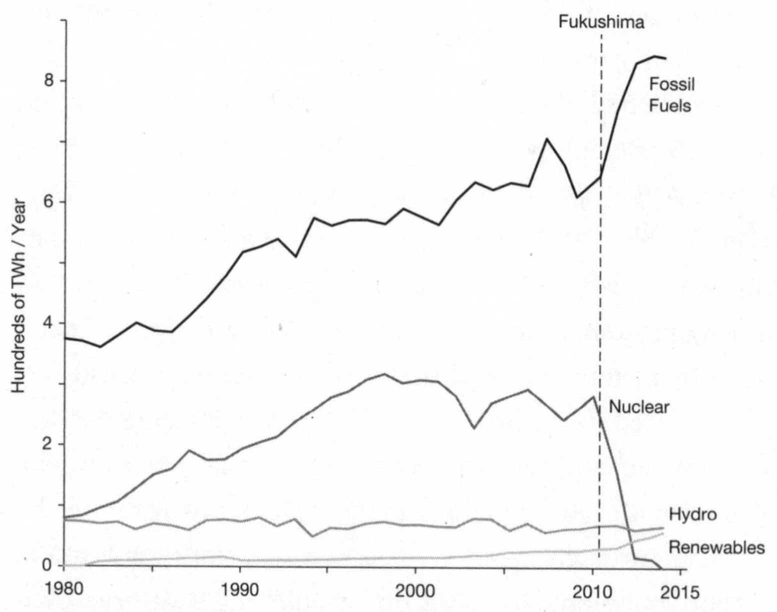

Figure 41. Japan's electricity production mix, 1980–2014. Renewables include wind, solar, geothermal, and biomass. *Data source*: US Energy Information Administration.

Japan is an extreme case. Six years after the Fukushima accident, it has forty-two usable nuclear reactors sitting idle with a capacity of around 40 GW, the equivalent of about ten Ringhals plants.[21] This massive electric capacity has been replaced by imported coal, oil, and gas, with new coal plants now planned.[22] Japan has blown through its carbon emission goals with little prospect of meeting its commitments. The country could save a lot of money and a lot of CO_2 by restarting its reactor fleet. By early 2018, more than six years after the earthquake/tsunami, about half of Japan's reactors had applied for permission to restart operations, but only five were back on the grid.

France had planned to cut its nuclear power production from 75 percent of all electricity to 50 percent by 2025 in response to post-Fukushima antinuclear activism. However, in late 2017, the French environment minister announced a delay in that action for a decade. The minister, a leading French environmentalist, declared that France's priority must be to first shut down coal and other fossil-fuel plants to reduce carbon emissions.[23] These statements were followed up by French president Emmanuel Macron, who argued that France would not follow Germany's example by phasing out nuclear power, because his priority was to cut carbon emissions and shut down polluting coal-fired production. Vive la France![24]

South Korea, one of the most successful nuclear power producers in recent years, has also flirted with a nuclear power shutdown recently. Decades ago, the country began by importing nuclear technology and then trained a generation of its own scientists and engineers to become self-reliant. With a standardized Korean design, it cranked out nuclear power plants very successfully and generated electricity at less than 5 cents/kWh.[25] But before South Korea's 2017 presidential election, Greenpeace and others heavily publicized a feature film about a nuclear power disaster, complete with a lot of explosions and deaths. This campaign moved public opinion to some extent, and a new president was elected who favored phasing out South Korea's nuclear power plants.[26]

The president halted construction on two nuclear power plants that had already received more than $1 billion in spending, and he appointed a citizens jury of almost five hundred people from all backgrounds to recommend future actions regarding nuclear power. In October 2017, the jury recommended, and the president accepted, restarting construction on the two reactors. However, it favored a longer-term policy of phasing out nuclear power. Six additional reactors that had been planned, but have not yet begun construction, remain canceled.[27] If South Korea sticks with its phaseout plan, nuclear power will be replaced with methane, costing billions of dollars a year and making it impossible for South Korea to meet its Paris climate commitments as older nuclear power plants retire around 2030.[28]

Even in Sweden, political winds are blowing against nuclear power. A new government took over that included the

Green Party in its coalition, and several reactors are being retired ahead of their useful lives.[29] If Sweden followed Germany and rapidly shut down its nuclear power stations, the results would be dramatic. According to a recent study, even with Sweden's extensive hydropower, the intermittent nature of wind and solar would force the country to install a large amount of excess capacity. To meet demand of about 150 terawatt-hours per year, wind and solar production of more than 400 TWh would be needed, along with grid upgrades. The cost of electricity would increase fivefold. Alternatively, the amount of new wind and solar could be cut in half—still greatly increasing electricity costs—by using fossil fuel as backup for the intermittent sources. But this would increase carbon emissions.[30]

A better solution would be to *increase* Sweden's nuclear power output to decarbonize its transportation fleet and export clean energy to northern Europe's grid to displace German and Polish coal.

CHAPTER TWELVE

Next-Generation Technology

MOST OF TODAY'S nuclear reactors are "second generation" (the first generation being early experimental reactors). They work, they are safe, and they are proven through long history. Any country that has good second-generation reactors—especially Japan and Germany—should certainly keep them running. Any country that has the ability to build out second-generation reactors in much larger numbers at low cost should certainly consider that option.

The Third Generation

Third-generation designs followed on the 1986 Chernobyl accident and claim greater safety in unusual circumstances, such as earthquakes or failures by human operators. They are "walk-away safe" in that the reactor will shut down safely, without danger of a meltdown, with no operator action needed for at least seventy-two hours. Principles such as gravity replace human intervention. By contrast, a Fukushima-type design relies on electric power to pump cooling water after an incident, and this approach failed when all the backup generators were swamped by an epic tsunami.

Several third-generation designs have entered operation successfully, but those under construction in Europe and the United States are in trouble. Designed to be simpler and cheaper to build, several of them have instead proven extremely expensive, with long delays and regulatory snafus.

In the United States, in particular, a nuclear power industry that had not built any new reactors in decades had a terrible time trying to get back into the game with designs that had never been built before. So far, exactly one new reactor (Watts Bar in Tennessee) has reached completion and entered service. But it was a second-generation design that had been suspended in the 1980s and—needing design changes to meet new regulations during its forty-two-year construction—ended up costing $12 billion.

The leading third-generation design in the United States was the Westinghouse AP1000, which not only was inherently safer but also required less material in construction and was simplified in order to be more economical. The design was approved by the Nuclear Regulatory Commission in 2005, and four reactors were to be built, but a decade of delays and cost overruns ensued. Regulations and designs kept changing, and antinuclear groups kept up a stream of objections and legal actions against the plants. Westinghouse lost so much money, almost $10 billion, that it filed for bankruptcy in 2017, threatening its parent company, Toshiba, as well. The two reactors in South Carolina were abandoned, leaving the remaining two, in Georgia, hanging by a thread. (China is moving forward with four AP1000s, although they are delayed by several years, and the first connected to the grid in 2018.)

The other big American nuclear power company is General Electric. With its Japanese partner, Hitachi, it has decades of reactor experience and a third-generation design, the Economic Simplified Boiling Water Reactor. In 2017 the NRC granted approval for construction of an ESBWR in Virginia. It took nine years for the NRC to certify the ESBWR design and several more years before approval of the Virginia plant. All of that is before a shovel ever hits the ground. A second ESBWR, planned for Mississippi, was abandoned after ten years, before construction.

The French nuclear power company AREVA has created a third-generation reactor called the EPR (originally the "European Pressurized Reactor"). One is being built in France and one in Finland, both taking about a decade in construction, years behind schedule and billions of dollars over budget. China is building two EPRs, which, although a couple of years behind schedule, are being built faster and cheaper than those in Europe. The first came online in 2018. Britain has started to build two EPRs as well, and in 2018 India ordered six.

Russia has one third-generation reactor in operation[1] and several more in construction or planned, including reactors for export to Finland and elsewhere.

Figure 42. Heads of the IAEA and United Arab Emirates nuclear company view construction of the Korean-designed APR1400 in 2013, five years before opening. *Photo*: International Atomic Energy Agency.

Finally, and perhaps most important because of cost considerations, South Korea has a third-generation reactor, the APR1400. One is in operation and others under construction, including the four reactors being built in the United Arab Emirates. The APR1400 is arguably emerging as the third-generation front-runner. The UAE plant is nearing on-schedule completion at a price of $4 to $4.5 billion/gigawatt of generating capacity—higher than South Korea's plants but still competitive in the world market.

The Fourth Generation

With the problems in building third-generation designs and the general malaise in the European and US nuclear power industry, dozens of new companies are developing new, fundamentally different designs. Several state-backed initiatives, and dozens of privately funded start-up companies with more than $1 billion of investment, have been developing designs for "fourth-generation" nuclear power plants.[2]

The regulatory system in the United States is geared to second- and third-generation light-water reactors. It needs to adapt rapidly to handle new and more radical designs. Currently, as MIT's Richard Lester writes, a company proposing a new fourth-generation design faces "the prospect of having to spend a billion dollars or more on an open-ended, all-or-nothing licensing process without any certainty of outcomes or even clear milestones along the way."[3] No wonder that most of the fourth-generation start-ups have moved to develop their designs outside the United States.

One of the most prominent of the fourth-generation companies, Terrapower, was cofounded by Bill Gates.[4] Its concept is a reactor that "breeds" plutonium from uranium and then uses the plutonium as fuel, all contained within the reactor. The concept is called a "traveling wave reactor," as the fission reaction gradually moves through the fuel over the lifetime of the reactor, several decades.[5]

If this or a similar concept panned out, we could see a large cylinder buried in the ground and used to produce electricity for many decades and then eventually dug up and replaced. Unlike previous "breeder" reactors, which require plutonium to be exported off-site for reprocessing, the whole thing would be self-contained—safe from weapons proliferation, from terrorist attack, from tsunamis, and so on. The company's hope is to produce electricity this way cheaper than from fossil fuels.

Terrapower found the regulatory environment in the United States problematic and signed agreements with China to develop the reactor there.[6] The first of these reactors will be built and operated in China, hopefully by 2025, and then they could potentially be exported around the world and make a significant difference in decarbonization.

Several companies are pursuing the concept of reactors in which the fuel is a liquid (a molten salt) rather than solid fuel rods, such as today's reactors use. Several forms of molten salt have been used as the medium for such nuclear fuel. An experimental reactor created by the US government in the 1960s proved that such a design could work, but eventually it was dropped in favor of sodium- and water-cooled reactors with solid fuel rods that are in use today. Molten salt designs are being developed by many start-ups, such as the Canadian company Terrestrial Energy, as well as within a state-funded program in China.

The Terrestrial Energy reactor is an example of a Small Modular Reactor (SMR). Other SMR designs, such as that of Oregon-based NuScale Power, are light-water reactors, or cooled by liquid metals, in addition to the molten salt designs. The idea of the SMR is a smaller reactor that is flexible enough to deploy close to where power is needed and can be deployed in groups to achieve any needed level of power capacity. Being small, the reactor can be built in a central location, minimizing expensive on-site work and standardizing design. For example, Terrestrial hopes to build a reactor within four years and produce electricity at a price cheaper than fossil fuel even without a carbon price added to the fossil fuel. Terrestrial and several other fourth-generation companies have received support from a US government program that lets them partner with government labs to test their ideas and bring them to fruition.

Another fourth-generation idea is to use thorium, as well as uranium, as the main fuel, sometimes also in a molten

Figure 43. Small Modular Reactor central-construction concept.
Graphic: Courtesy of NuScale Power, copyright © 2007 or later by NuScale Power, LLC.

salt reactor. Inside the reactor the thorium is converted to uranium-233, which fissions. Thorium has several advantages over uranium and was explored by the US government in the 1960s but ultimately passed over, partly because it was less useful in nuclear weapons production.[7] India hopes to use thorium as its main nuclear fuel in the coming decades, and several fourth-generation companies are trying to commercialize it.

The company ThorCon is among several trying to develop liquid-fueled thorium reactors.[8] Like the Terrestrial

Energy and NuScale reactors, the ThorCon reactor would be produced centrally and shipped to the site, which could be the shallow seabed just offshore. It contains two "cans" of fuel, one producing power, the other cooling off. Each would be replaced every few years so that the reactor keeps producing power during periods when servicing and refueling are taking place at a central location. Like other molten salt reactors, theirs would be walk-away safe, as fuel would drain out and cool down without intervention in an emergency.

The leaders of ThorCon come from a shipbuilding background, and the key to success for them is mass-producing units centrally using shipyard methods, to bring down costs. The reactors would be built in "blocks" that can be barged to and from remote sites, with standardized design and components that are theoretically cheaper than those in coal power plants. If successful, ThorCon plans to produce CO_2-free electricity for 3–5 cents/kWh.

ThorCon is in discussions with the Indonesian government to potentially install ThorCon reactors there—probably built in shipyards in Asia or Europe—to fuel Indonesia's expected rapid growth in electricity demand. Eventually, the company aspires to produce one hundred plants annually, each with 1 GW of capacity, just as several large shipyards can produce one hundred big ships per year using similar block-construction methods. That would be the equivalent of a new Ringhals plant every two weeks—an ambitious plan for rapid decarbonization.

Figure 44. Several shipyards like this one in South Korea in 2015 can together produce one hundred complex ships per year, each with as much steel as a ThorCon 1 GW unit. *Photo*: SeongJoon Cho / Getty Images.

Another idea for using shipyards to build nuclear power plants comes out of MIT. The proposal of a group there is to combine two well-established technologies, building nuclear power plants and building offshore oil-drilling platforms. The resulting nuclear power plant would sit 8–10 miles offshore, far enough to avoid siting issues and tsunamis or earthquakes, but within territorial waters. It would send up to a gigawatt of power through an underwater cable to the grid onshore. Like an oil platform, the top levels of the cylindrical structure would hold quarters for a work crew that would rotate in for a month or two at a time and a heliport to land them. The nuclear reactor would sit below water level, in containment structures, and make use of seawater for cooling and for passive emergency backup cooling. Costs would be low because of the central construction, shipyard methods, and the use of steel with minimal concrete. If circumstances changed, or at the end of its sixty-year life, the plant would be towed somewhere else.[9]

Recent variants of the MIT concept would use a larger, shiplike platform with two large reactors of a type already licensed and built, such as South Korea's APR1400. Located either in a harbor setting, such as at an existing nuclear power plant site, or 10 miles offshore, the floating unit would solve siting problems in places like India where land near the shore is crowded. In terms of safety, the ocean is the ideal location for a light-water reactor, given that all past major accidents involve failure of cooling. In terms of proliferation, the offshore design means that the entire fuel cycle would never touch shore, with no temptation for countries to divert material to a secret weapons program. Most importantly, by replacing construction with manufacturing, at an advanced shipyard that can churn out dozens of identical units, the floating reactor could achieve breakthrough low cost, cheaper than any fossil alternative. Because of the centralized, factorylike production system, the operation would be scalable to the rapid growth the world needs—something much harder to achieve with site-specific construction in each location.[10]

One thing many of the fourth-generation companies have in common is a decision to push forward outside the United States, whether in China, Canada, or Indonesia. If the US government wants to reclaim leadership of the world nuclear power industry and make the United States a leader for these new technologies, then government support of technological progress must be stepped up considerably. The program giving access to government labs and collaboration with scientists is a good start. But the whole regulatory system for nuclear power is not yet well adapted for these new designs. There is a strong national interest in developing these technologies that may prove not only vital for saving humanity from climate change but also commercially productive as contributors to US economic health.

From a political marketing perspective, fourth-generation nuclear power has several advantages over existing designs. Fears about nuclear power, however ill founded, might be mitigated with the reassurance that fourth-generation designs are "even safer" (than safe) and all-around better. Furthermore, psychological research shows that people prefer products that compare favorably with similar products.[11] (That is, fourth-generation nuclear power designs are superior to existing ones.) Another political advantage is that fourth-generation designs have real potential for cheaper power. And people do like shiny new versions of technologies, such as cell phones. There is strong bipartisan backing in the US Congress for government support of fourth-generation research and development (R&D).[12]

However, a danger in fourth-generation designs is that the public and politicians may think we need to wait a decade or two for these new designs instead of using what we already have. That would be like refusing to buy a smartphone because you believe that brain implants will do the job more elegantly in another decade or two. And the claims of greater safety may make current nuclear power designs seem unsafe—otherwise, why would we need safer ones? But we have extremely safe designs already and can move forward with nuclear power expansion at the same time we develop the fourth generation to carry the next stage of expansion.

Fusion

Further in the future are more radical plans such as a fusion reactor (fusing small atoms like hydrogen, as the sun does, rather than splitting large ones like uranium). An international collaboration of thirty-five countries, based in France, is building a $20 billion 500 MW fusion reactor with 10 million parts, which its builders call "fantastically complex," and hopes to begin operation in 2035.[13] At the other end of the size spectrum, a start-up company near Vancouver, General Fusion, has raised $100 million in investment for its small fusion reactor; it hopes to have a prototype in the next decade.[14]

Fusion power and other energy breakthroughs should continue to be pursued, but meanwhile, in the medium term, fourth-generation fission designs could potentially take off and supply much of the needed electricity as the world economy grows in the middle decades of the century. We can't count on this alone to work, but we should pursue it vigorously as a promising possibility.[15]

Geoengineering

A different and nonnuclear kind of new technology that deserves mention is "geoengineering." Reflective geoengineering refers to seeding the stratosphere with tiny particles of sulfur (or another aerosol) to reflect back some sunlight and slow the warming of the planet.[16] We know that such a process works in cooling the planet because we have seen modest cooling after major volcanic eruptions, which spew a lot of sulfur into the high atmosphere. (Ironically, even particulates from coal burning have some cooling effect, although more than offset by the warming effect of CO_2.) We authors favor research into this possible intervention, which might buy the world a bit of time later.[17] However, it is important to understand its limits.

Reflective geoengineering does not reverse the effects of greenhouse gases; it overlays them with another process

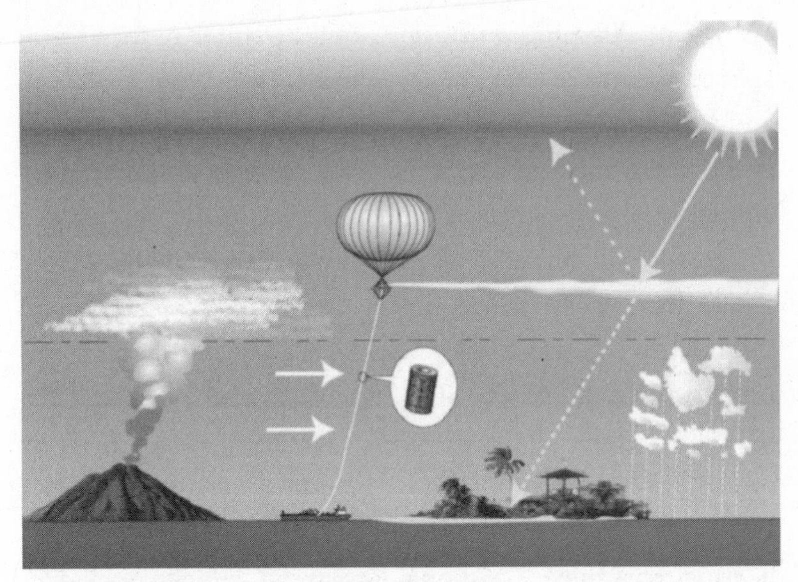

Figure 45. Reflective geoengineering seeks to mimic the effect of volcanic eruptions in cooling the atmosphere, as in this 2011 concept. *Graphic*: Hugh Hunt via Wikimedia Commons (CC Attribution-Share Alike 3.0 Unported).

mostly moving in the other direction.[18] But the two processes are not opposites. They work differently according to time of day and season of the year. They affect the ocean entirely differently, with CO_2 acidifying the water, while sunlight reflection does nothing to correct that change. Because of these differences, reflective geoengineering may slow the warming of the planet overall but might make things worse in some regions. It might change weather patterns in ways that increase drought in some places and possibly floods elsewhere. Anytime we stopped putting sulfur in the stratosphere, the entire effect of greenhouse gases would catch up and temperature would jump disastrously. (Sudden change is the worst form of climate change, as ecosystems cannot adapt.)

Because of these problems, reflective geoengineering would have to be started only after careful research and then phased in very slowly over several decades and eventually phased out slowly over several decades. Although potentially a useful tool in the future, it is in no way a quick fix or a means to rapidly stop temperature change before we lose control of the planet's climate. As a report on geoengineering from the US National Research Council concluded, "There is no substitute for dramatic reductions in greenhouse gas emissions" to mitigate climate change.[19]

A different idea also sometimes referred to as geoengineering is to suck CO_2 directly out of the atmosphere. It is entirely benign and positive, but so far way too expensive to be practical. As technology develops in the coming decades, we will need to pursue methods to remove CO_2 from the atmosphere to bring concentrations down to safe levels.

In the shorter term, "carbon capture and sequestration" (CCS) refers to sucking CO_2 out of power plant exhausts, before it enters the atmosphere, which is easier because the CO_2 is more concentrated. Although still expensive, CCS is a potentially important measure that should be pursued but does not change the need to rapidly decarbonize. In 2017 a multibillion-dollar CCS "clean coal" effort was abandoned, but a methane CCS design shows promise.[20] Methane power plants that did not emit (much) CO_2 could contribute substantially to decarbonization, although the problem of unburned methane leaks would remain.

CHAPTER THIRTEEN

China, Russia, India

IF THE WORLD is to rapidly decarbonize, that process will depend heavily on what happens in three large countries. China is by far the largest emitter of CO_2 but also building more nuclear reactors than anyone else, as well as more solar and wind power. Russia is a fossil-fuel giant exporting to the world, but Russia also exports more nuclear power plants than anyone else. India is the world's third-largest CO_2 emitter, though a distant third after China and the United States, with a coal-based generating system that will need to grow rapidly to keep up with fast-growing energy demand for a huge, poor population.

Much of the discussion of climate-change solutions has revolved around the Western industrialized democracies and especially the United States. It's true that those countries created the problem with their historical use of fossil fuel and also that they use a lot of energy and still add a lot to CO_2 pollution. But these countries have some advantages not shared by the poorer parts of the world. Using technology, they have made energy use more and more efficient in recent decades, so that their carbon emissions are slowly declining—not dropping fast enough, but not rising either. And the rich countries are rich enough that if they decide to spend more for cleaner energy, they can do so without courting disaster. Furthermore, the United States can largely drop coal because of its cheap fracked methane.

The world's poorer countries face greater constraints. The farmers in India who desperately want electricity, so their children have light to study by at night, cannot afford to pay a premium rate for it. Russia's economy depends on oil and gas exports and would collapse if those suddenly stopped with no comparable income stream to replace them. China's rapid economic growth would grind to a halt, possibly with severe consequences for its social stability, if energy supplies could not keep up. In all three cases, affordable energy is not a luxury but an economic and political necessity. In each case, however, a rapid expansion of nuclear power could fill the gap (in Chinese and Indian energy consumption and in Russian energy exports). It helps that all three countries have well-developed civilian nuclear industries as well as already owning nuclear weapons, making proliferation a nonissue.

People in Western democracies might have the impression that nuclear power is dying out, but this is not true worldwide. Some 449 reactors in thirty-one countries are in operation, producing 11 percent of the world's electricity, second only to hydropower as a clean energy source.[1] In 2018 of the 53 new reactors under construction in fifteen countries, the majority were in China, India, and Russia.

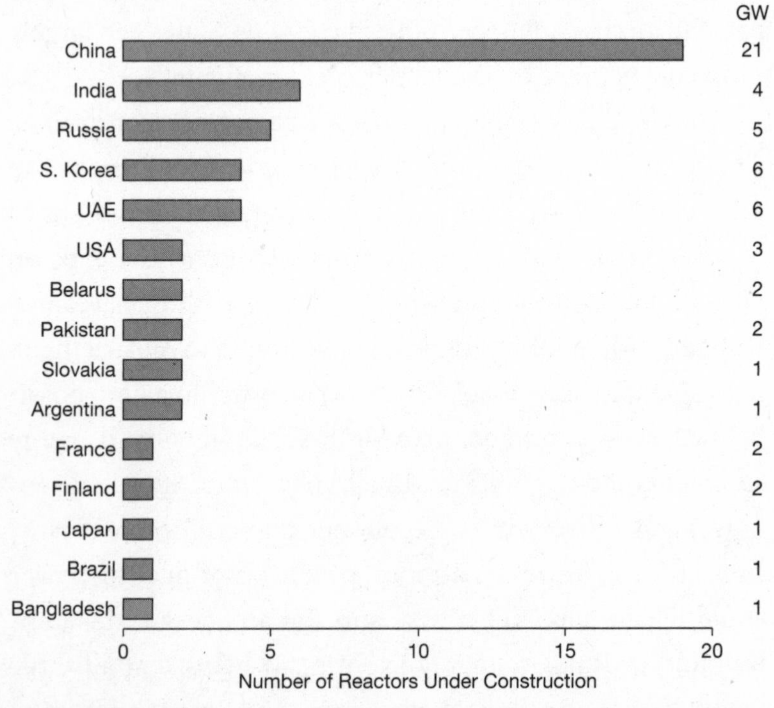

Figure 46. China leads in nuclear power construction. *Data source*: World Nuclear Association.

China

China is key to the climate crisis. Its breathtaking economic growth over thirty years has not only raised Chinese citizens' standard of living dramatically but also made China the industrial workshop of the world. China has taken the lead in producing solar and wind equipment ever more cheaply for use worldwide. At the same time, China keeps burning huge amounts of coal, and its emissions remain dangerously high, with smog choking its major cities. Worse yet, one new method for reducing the smog involves using coal in distant rural locations to create gas that can be burned more cleanly in the cities, but at the cost of even higher total greenhouse emissions from the gasification process.[2] Chinese companies are also major players in building coal plants in other developing countries, with hundreds of gigawatts of new capacity planned.[3] Solving the world's climate problem without solving the Chinese coal problem is flat-out impossible. The Chinese government says its CO_2 emissions will peak around 2030, and many experts think this will happen even sooner. But, as we saw in Chapter 1, merely flattening out high emissions at their current levels does not solve the climate problem.

What, then, can replace coal in China, and quickly? The country is building vast solar and wind farms as fast as it can, but these just chase after rising demand while the existing coal plants churn on. Energy use is not going to suddenly decrease. The only practical answer is to do what Sweden did, but on a bigger scale. One of us (Qvist) recently visited China to recommend just this program to the leadership there. The response was that, among other things, China lacked the number of trained and experienced experts to be able to expand that far and fast. However, this problem can be overcome.

First, China can import expertise from elsewhere, especially from places like Germany, where nuclear power plants are being shuttered. China can also develop stronger cooperation with the United States to build human resource capacity.[4]

Second, China can standardize and replicate. In both France and South Korea, a key to success was picking a standard reactor design and building it over and over, with the same team of experts moving on from one to the next. South Korea created the Korean Standard Nuclear Plant (KSNP). China has not settled on a standard design but is still building a number of different designs, and it is arguably a large-enough economy to pursue a variety of designs at scale simultaneously. But by choosing a small number of main technology tracks, China could quickly get really good at building reactors and begin to produce them more like Boeing jets rolling off an assembly line and less like bridges uniquely designed for a particular site.

The keys to low-cost nuclear power, according to a recent review of the history of the industry, are standardized design applied repeatedly, government support, and building multiple reactors at each site.[5] China could readily achieve all three. It need not follow the US model of multiple designs built and operated by a variety of private companies. While the United States has seen an escalating cost curve for new reactors, the trend in South Korea has been downward, as lower costs followed from greater experience. Already, China is able to build a 1 GW plant for $2 billion and generate nuclear electricity for around 3–6 cents/kWh, cheaper than any source other than hydropower.[6]

China is currently operating thirty-seven reactors and building nineteen more. Its main design for new builds is the Westinghouse AP1000 and Chinese variants of it, notably the CAP1400.[7] But China is also building EPRs and several

Figure 47. China's wide variety of reactor types includes the important third-generation CAP1400, here under construction in 2014. *Photo*: Conleth Brady / International Atomic Energy Agency.

Chinese designs, including both large reactors and Small Modular Reactors, as well as Russian and Canadian designs, very small reactors, and floating reactors. China hopes to transition, in several decades, to "fast" neutron reactors and fourth-generation technology.[8] China is spending $3 billion to build two prototype molten salt reactors by 2020.[9]

The only problem with China's shotgun approach using so many reactor designs is that coal, meanwhile, continues to burn at ruinous rates. China could focus on mass production of existing light-water reactor technology and transition quickly away from coal while maintaining a robust R&D budget to seriously prototype and commercialize one or more fourth-generation technologies within the next decade. The AP1000 (and its Chinese follow-on, the CAP1400) reactor might be one of the workhorses replacing coal plants in the near term. The first Chinese AP1000 began electricity generation in 2018, with the second set to begin shortly after.[10] If successful, many more could feasibly follow in short order. The Chinese-designed Hualong One, another third-generation reactor, is to follow a year later. If scaling up these designs proved problematic, China could look to the Korean KSNP, bringing in Korean engineers who may be out of work if South Korea stops building nuclear power plants and who have recent export experience from the four nuclear power plants they are building in the UAE. By paying handsomely for technology transfer, and creating a crash program to replicate the proven design widely across China—rather than spend years to prove out a new design or variant—China could make a very fast Swedish-style transition from fossil fuel.

A small set of standardized, proven designs replicated on a large scale across the country would provide China a path out of coal, arguably doing more to fix the climate crisis than any other single action in the world. It would also boost the popularity of the Communist Party leadership and President Xi Jinping by making city air breathable again.

As a rollout along these lines across China reached maturity, China could export nuclear power plants to the poor countries of the world. China's specialty recently has been building infrastructure, such as ports, railroads, roads, and indeed a large number of coal plants across Asia and Africa, as part of its so-called One Belt, One Road initiative—sometimes called the New Silk Road. China often operates what it builds; for instance, the person sweeping up a railroad station in Ethiopia may well be a Chinese citizen. What works for transportation infrastructure might work for energy infrastructure. China has already begun to export nuclear power plants along with other infrastructure to developing countries. When China mass-produces nuclear power plants that can safely produce cheap electricity, it could sell a lot of them to energy-hungry countries, although there is no guarantee of success.[11] As Japan, South Korea, North America, and Europe drop out of the game—if indeed that trend continues—China could clean up.

Russia

Russia is a bit more complicated. Domestically, it could in theory replace coal (about 20 percent of generation) with methane and nuclear power with some good effects. But Russia can get more money for its methane by exporting it, so in practice only nuclear power can replace coal. Russia already operates thirty-five nuclear reactors,[12] accounting for almost 20 percent of electricity generation, similar to the United States. But unlike the United States, Russia is building a new reactor each year to replace retirements and raise nuclear power's share of electricity generation to 50 percent by midcentury and three-quarters by century's end.[13]

The unique potential of Russia lies in its status as the world's current leader in exporting nuclear power plants, with about 60 percent of world exports.[14] Russia specializes in exporting turnkey plants that it will design, build, and operate. Electricity prices are projected at a competitive 5–6 cents/kWh. Current export orders include thirty-four reactors in thirteen countries at a total cost of about $300 billion.[15] In 2017 the government decided to reduce financial support to the national nuclear power company, Rosatom, making exports all the more important in the company's future. Russian export reactors are already finished and operating in China, India, and Iran. More are contracted in Bangladesh, Vietnam, Turkey, and Finland, and orders have been placed by Belarus, Bangladesh, Turkey, Hungary, and

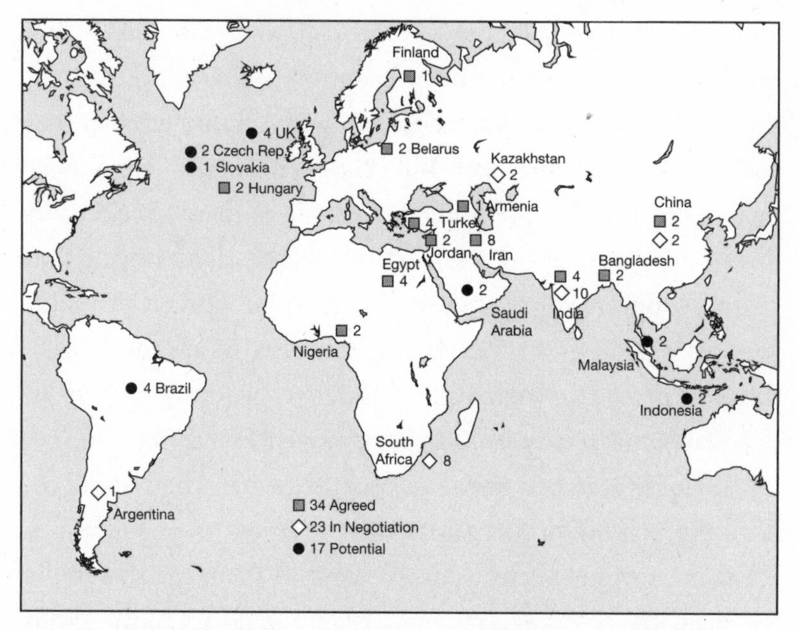

Figure 48. Russian nuclear power export projects, 2016. *Data source*: Rosatom, *Performance of State Atomic Energy Corporation Rosatom in 2016: Public Annual Report, 2017* (Moscow: Rosatom, 2017), 80.

Egypt.[16] Broad agreements have been reached, but without specifics, in other countries in Asia, Africa, and Latin America. Many more energy-hungry countries contemplating new coal plants could, instead, consider nuclear power plants built and operated by Russia.

Russia has its own version of fourth-generation technology, a program called Breakthrough, to develop fast reactors with closed fuel cycles. (Fast reactors use neutrons that are not slowed down as in conventional "thermal" reactors and include the "breeder" reactors that create their own fuel, plutonium, as they burn uranium.) The Breakthrough designs can run on nuclear waste, with characteristics that somewhat resemble the Bill Gates–backed Terrapower design discussed in Chapter 12. Russia is firmly placing its future energy bets on nuclear power and does not have a strong program supporting wind or solar power as major sources of energy in the coming decades. With the Breakthrough program, the country hopes to generate almost 50 percent of electricity from nuclear power by 2050.

Because these new concepts recycle the fuel, they would reduce the volume of fuel and waste far below the already low levels in previous nuclear power plants. A plant the size of the Swedish Ringhals nuclear power plant would use only 3 tons of fuel and produce 3 tons of waste per year.[17] That compares with the Jänschwalde coal plant at about 50,000 tons of fuel and waste per *day*! To put this in another perspective, the average American uses on the order of one gigawatt-hour of electricity in his or her lifetime. With the Breakthrough (or a similar fourth-generation) design, this lifetime supply would require one-quarter of one pound of fuel and generate the equivalent amount of waste. That's the weight of the beef in a single hamburger for a lifetime of American-level electricity use.

The Breakthrough program is proceeding at nine coordinated research centers with strong government support. As with other fourth-generation projects elsewhere, the goals are to create inherently safe reactors with closed fuel cycles (little waste), low cost, and prevention of proliferation. The plan is for a tenfold expansion of Russian nuclear power generation by the end of the century using the new designs.

Russia has a head start on the rest of the world when it comes to these designs. It has been operating what in many ways amounts to a fourth-generation plant since 1981, producing electricity at competitive rates on the Russian grid. The BN-600 reactor, in central Russia, is living proof that these technologies can work commercially in the real world, not just in theory. Its bigger sister, the BN-800, another fourth-generation reactor, has recently entered commercial operation and is supplying the Russian electric grid. These designs will culminate in a larger reactor, the BN-1200,

Figure 49. Russia's four-reactor Beloyarsk power plant includes the BN-600 and new BN-800. *Photo*: Rosenergoatom.ru.

planned for construction at several sites in the next twenty years. The head of Rosatom said in 2017 of the Breakthrough technology that "today we are leading in this field. It's necessary to make this leadership absolute and to deprive our competitors of their hopes of overcoming the gap in the technological race." (Listen up, US government.)

Russia's first small floating nuclear power plant is scheduled to start producing in 2019 in the remote Northeast of Siberia. This ship and one coming soon in China have the mission of providing heat and electricity in remote areas. But a more significant use of floating power plants, in theory, is for export to any number of coastal locations in the world—and most of the world's big cities *are* coastal—to sit offshore and supply competitive CO_2-free electricity. The floating plant reduces problems with siting, as land tends to be expensive near big cities and populations may have "Not in my back yard" attitudes. When the plant is ready to decommission, it can be towed back to its home country. The American fourth-generation company ThorCon has a similar concept, as does a group at MIT (both mentioned in Chapter 12), but the Russians are closer to realizing the concept with a more conventional light-water reactor similar to those already used in their nuclear icebreaker fleet.

In recent years, Russia has seemed to show indecision about its energy future. It has a lot of cheap methane, which could supply electricity. (Russia is also expanding its hydropower, which currently supplies a bit less than 20 percent of electricity.) But the methane can make more money being exported to Europe than generating electricity at home. If Russia succeeds in sharply increasing its nuclear power capacity, it could export much more gas to Germany, which could use it to replace coal. This would moderate the negative climate effects of Germany's decision to phase out nuclear power. Such gas exports, however, are currently tangled in sanctions and geopolitics because of the war in Ukraine.

If Russia mastered commercialization and mass production of fourth-generation Breakthrough technology ahead of the rest of the world, it could significantly and quickly expand the world's clean electricity supply. This would transform Russia's role in the world climate picture from part of the problem (massive fossil-fuel producer and exporter) to part of the solution.

India

India expects, and deserves, large and rapid increases in its electricity use in the coming few decades. Currently, it meets this growing demand primarily with coal and is the second largest consumer of coal in the world, albeit far smaller than China. Far from phasing out coal, India continues to increase coal-fired generation capacity, although at a slowing rate.[18] India's large, ambitious, and well-publicized foray into solar power, though laudable, will at best slow the growth of coal power.[19] Something else is needed if India is to phase out coal and meet its growing demand without large CO_2 emissions. It is hard to see what else that "something" could be if not nuclear power.

India is an unusual case when it comes to nuclear power. It did not sign the Non-Proliferation Treaty in 1970, and it set off nuclear weapons test explosions in 1974 and 1998. India's arsenal of nuclear weapons is estimated at about 125 warheads.[20] Because of its status outside the NPT, India was subject to international restrictions during most of the time its civilian nuclear power industry was developing. After 2008 the international community came to terms with India's nuclear weapons status and began making deals for nuclear fuel and technology trade with India, under specific arrangements for IAEA safeguards. The plutonium India uses for weapons comes from plutonium production reactors that do not produce commercial electricity and are not under IAEA supervision.

Since the 1950s, India has pursued a three-stage nuclear power program: first, set up Pressurized Heavy Water Reactors (PHWRs); later build a series of fast breeder reactors that use a mix of uranium and plutonium as fuel; and, eventually, create a fleet of thorium-fuel-based reactors regenerating most of their fuel with the fissile material produced in the fast reactors. India's long-term target of developing thorium-fueled reactors reflects the fact that the country has little uranium but plenty of thorium, which can change into fissile uranium in a reactor.

The unusual choice of PHWRs for the first stage of nuclear power expansion is primarily due to international restrictions on enriched uranium.[21] The first and second stages of the nuclear power expansion program are technically in operation, but at a surprisingly small scale. In 2018 India was putting into operation the 500 MW "Prototype Fast Breeder Reactor," which is its commercial prototype plant for fourth-generation fast-reactor technology that makes up the second stage of the program. If India is to decarbonize, it is imperative that the enormous efforts put into the PFBR are followed up by commercial fast-reactor units in the near future.

These aspects of Indian nuclear power put it on a self-sufficient path but one that does not mesh very well with the rest of the world. India plans to build new nuclear power plants but not very many of them, and it plans to import some plants but not very many of those, either. This would keep the Indian economy firmly in the fossil-fuel camp for decades to come—decades of anticipated rapid growth of electricity demand for a huge population.

A viable route for India to decarbonize is to accelerate its build-out of available reactor technology (PHWRs and, if it is able to source them from the international market, light-water reactors). In 2018 India took a big step in this direction by signing an agreement to build six EPR reactors totaling 10 GW.[22] India can also move toward commercialization of its indigenous advanced-reactor research programs. Essentially, it would follow its three-stage program but on a vastly larger scale and faster timeline. India also has much work to do in modernizing its electric grid. On the positive side, the country has tremendous technology assets, especially in "human capital" (an educated workforce), and a great deal of experience in the nuclear field. India may find its own way to put together its unique capabilities to create a climate-friendly development path.

CHAPTER FOURTEEN

Pricing Carbon Pollution

SWEDEN DECARBONIZED ITS economy not just by building reactors. It also changed the economics of energy across the board. Sweden not only has the world's highest use of nuclear power per person; it also has the world's highest price for carbon pollution. This carbon price is sometimes called a "carbon tax" or "carbon fee."

When people put CO_2 into the atmosphere, they create a cost to society by accelerating climate change. The air pollution from burning fossil fuels also comes at a cost, since the tiny particles in smoke cause cancer and other terrible diseases. These costs can be estimated, in theory. For instance, one recent study estimated the health effects of burning fossil fuels in the United States (not even including the climate effects) at about $900 billion a year. Including those costs in the price of electricity would at least double the price per kilowatt-hour.[1] Until now, in most of the world—but not in Sweden—polluting with carbon and other emissions has been free, with no consequences to the person dumping CO_2. The consequences are instead shared by everyone in the society, or by the world in the case of climate change, and by future generations.

We don't handle other types of pollution this way. For example, when people flush their toilets, we no longer just let the sewage run down the streets and into nearby waterways, with the costs in disease borne by others in the future. Rather, we install sewers and sewage treatment plants, and then we charge a sewer tax or fee based on how much water a household uses. Similarly, we charge a fee to dump garbage at a landfill or to pollute the air with smog from a car (charged by way of a gasoline tax and mandatory emission inspections).

All these are examples of a bigger class of problems in which the actions of an individual create costs for a whole group. Overfishing, tax evasion, and international alliance commitments are other examples of this type of problem. The solution is some form of governance that makes people pay the costs of their actions rather than passing them along to society. The Law of the Sea (for overfishing), the Internal Revenue Service (for tax evasion), and NATO defense spending targets (for alliance commitments) all seek to accomplish this task.

When it comes to dumping CO_2 in our atmosphere, however, these governance mechanisms are missing. Not only can one individual dump for free, but entire countries can dump vast quantities with the bill picked up by the other countries and future generations.

A carbon price can solve that problem, like a sewer tax does. By charging for the "externality" of the costs of carbon pollution, it provides an incentive to pollute less. For example, economist Gregory Mankiw notes that he could reduce his carbon pollution by buying a more fuel-efficient car, adjusting his thermostat, buying locally produced food, and so forth. One way to induce him to do these things is by moral persuasion, such as President Jimmy Carter used by wearing a sweater in a colder White House in the 1979 energy crisis. But Mankiw considers this approach "unrealistic," and it certainly has not solved the problem in the years since 1979. A second approach is government regulation, such as fuel-efficiency standards imposed on car manufacturers. But regulations are complex and have limited reach (for instance, changing how efficient the car is but not how far or fast people drive). They can even backfire, as when fuel-efficiency standards for cars contributed to an industry switch to SUVs and light trucks. The third approach is to charge a fee for carbon emissions, which broadly shifts incentives across the whole economy without requiring government to regulate every little action and decision.[2] Professor Mankiw can save money by using energy more

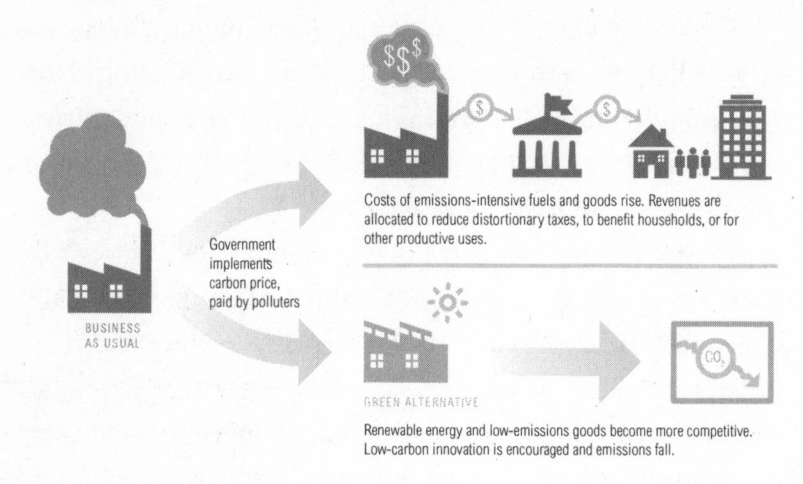

Figure 50. Carbon pricing concept. *Graphic*: Courtesy of World Resources Institute. www.wri.org/carbonpricing.

efficiently, and large-scale power sources that do not pollute, including nuclear power, gain a competitive advantage compared with fossil fuels (which must now carry their full costs).

Economists tend to love this approach because it is efficient and allows market forces to do the work. Rather than government (or privately enforced norms of behavior) dictating how much energy you use and what you do with it, the carbon price lets you do what you want but provides an incentive by charging you for putting CO_2 into the atmosphere. A survey of economists found that 90 percent favored a carbon tax rather than direct regulation as a way to reduce carbon emissions.[3]

Setting the right price is difficult. One approach is to try to calculate the future costs of a ton of carbon dumped in the atmosphere today. We know that those costs are not zero, but nobody claims to understand exactly how climate change will play out and what economic and social effects it will have. In 2013 the Obama administration published estimates of the "social cost of carbon," based on the increased likelihood of floods, reduced agricultural productivity, and so on. The result ranged from $11/ton to $123/ton, depending on assumptions—above all, how much future outcomes should be "discounted."

Discounting is controversial. A dollar in the future is worth less than a dollar today. The dollar today can be invested productively to become worth more than a dollar next year. The discount rate, which might typically be anywhere from 4 percent to 10 percent or higher, is the rate by which economists devalue outcomes year by year looking into the future. Critics say the discount rate literally devalues the future—at 10 percent discount, things forty years from now are valued below 2 percent of the same things today—implying that we don't care much about our grandchildren. But there is a legitimate debate about how much to spend today on our grandchildren's future well-being—for example, by reducing carbon emissions—versus how much to invest productively today and give our grandchildren the resulting pile of money to deal with problems themselves, such as by building seawalls.[4]

That way of thinking, however, is more applicable to minor climate changes and less so to catastrophic tipping points in which our grandchildren might not be able to cope at any price. Since linear increases in emissions lead to accelerating climate impacts, and eventually to an uninhabitable planet that cannot be fixed for any amount of money, the sensible and economical approach is to act now to halt emissions. Passing along to our grandchildren a big pile of money along with a broken global ecosystem does not make sense.

William Nordhaus, winner of the 2018 Nobel Prize in Economics, has explored the question of what price to charge for carbon pollution, given that too high a price leaves future generations with less money to fight climate effects in their own way, and too low a price leaves them with climate effects that will be more expensive to fight. To simplify Nordhaus's estimates, a low price of $11/ton of CO_2 worldwide would result in global warming of around 3.5°C by the end of the century. A price closer to $50/ton would more likely meet the 2°C target, if everyone in the world participated.[5] Nordhaus proposes that $25/ton in 2020, rising to $160/ton by 2050, would result in temperature rise of 2.5°C by the end of the century.[6] In our view, a more vigorous approach would be wiser and safer, given that nobody knows how well carbon pricing will actually reduce emissions or the ultimate effects of a 2.5°C temperature rise even if that goal were met.

Many companies have begun building a carbon price into their internal planning assumptions as they contemplate long-term investments. A survey of twenty-one US electric utilities in 2012 found that sixteen assumed a future carbon price, averaging almost $25/ton for 2020.[7] One report gives a spectrum of internal prices applied to operations in companies such as Microsoft ($6–$7/ton), Disney ($10–$20), Google ($14), and Exxon Mobil ($60–$80), among others.[8]

To date, all the efforts in individual companies, countries, and the European Union have been sporadic.[9] Only 12 percent of the world economy is covered by any form of carbon pricing. In the United States, in the absence of national carbon pricing, a number of individual states are exploring state-level carbon pricing.[10] These could work, but are tricky to implement because of the integration of the US economy across state lines. Ideally, a carbon price should apply to the whole world, but no practical mechanism exists to set such a price and implement it. When only some countries or regions charge for carbon pollution, there is an incentive for dirty industries to simply move to places where they can compete for free. Then the places with a carbon price have to impose a tariff (border adjustment) to account for the difference, but this is a difficult system to implement.

Sweden has had a carbon tax since 1991, currently the world's highest at more than $150/ton. Sweden has reduced emissions by 25 percent in that period (that is, *after* the big drop in emissions in the 1980s with the build-out of nuclear power). The carbon tax has driven shifts across the economy, notably in the use of biomass (forest and forest-product waste) instead of fossil fuels for district heating in cities (and heat pumps using clean electricity for houses without district heating). It has also spurred Swedish companies to create carbon-saving innovative technologies that find markets around the world, boosting the Swedish economy.[11]

For most of its history, the Swedish carbon tax applied at full force to households and services, and at a much lower rate—about one-third and as low as zero in some cases—to industries either covered by the EU Emissions Trading System (ETS) or deemed to be at risk of shifting production to countries without carbon pricing.

Germany does not have a carbon price. Its coal-powered electricity, with only the minor upcharge of the EU's ETS, goes onto the northern European grid and competes with Sweden's clean electricity. You might think that Germany is only burning so much coal to meet its own needs as its nuclear power plants shut down, but in fact in 2015 (the latest data available), Germany exported 48 TWh of electricity,[12] about the output of two massive coal plants such as Jänschwalde. The competition from coal-powered, carbon-tax-free German electricity has undermined the profitability of Sweden's nuclear power plants.

Collecting a carbon tax/fee is not too difficult because it can be charged at the first point where fossil fuels enter an economy—at the wellhead where fossil fuels are produced or the port where they are imported. Instead of charging each car driver for the CO_2 that comes out of his or her tail-pipe, the government can charge the fee on the crude oil before it is even turned into gasoline.

To gain support for a carbon fee, some advocates favor using all the revenue to lower other taxes or send rebate checks to citizens. This revenue-neutral concept goes over well with conservatives concerned that a carbon tax will be just another way for liberals to raise revenue and expand the scope of government. A US organization, the Citizens Climate Lobby, is pushing such a proposal, called Carbon Fee and Dividend. This approach sets a price of $15/ton, rising by $10 each year, and refunds all the revenue back to the citizens in the form of a regular dividend check. The dividend for a typical family of four would rise from $50/month to about $400/month over a twenty-year period.

The Canadian province of British Columbia has successfully implemented a revenue-neutral carbon fee.[13] The tax began in 2008 at about $9/ton of CO_2 and rose to about $27 by 2012.[14] The tax was revenue neutral because other taxes were cut by the same amount that the carbon tax brought in. Corporate taxes dropped from 12 percent to 10 percent. Emissions fell modestly, and the province's economy

Figure 51. British Columbia's carbon tax has not hurt its economy, including Vancouver's. *Photo*: Ajith Rajeswari via Wikimedia Commons (CC Attribution 3.0).

performed well, with most businesses coming around to supporting the tax. Now the whole of Canada is planning a carbon price, to start around $9 in 2018 and rise to about $45 by 2022.

On the whole, carbon pricing has great potential to help slow climate change. It takes effect quickly and changes behavior across the whole economy without needing to tinker with each wind turbine or electric-car charging station in an effort to engineer similar outcomes.

The oil shocks of the 1970s illustrate these economic effects. They quickly led to energy-efficiency measures such as driving less and lowering thermostats. They spurred innovations such as more fuel-efficient vehicles and the invention of the compact fluorescent lightbulb in 1976 (though it was not commercialized until later). And the oil shocks led to policy changes such as Sweden's decision to build nuclear power plants. A serious price on carbon pollution would have similar beneficial effects.

However, carbon pricing has proven politically challenging and needs to be better understood by the public. And its effects are not easy to predict with certainty. People often adjust to prices after an initial shock wears off, so a given carbon price might have less effect over time, as happened in British Columbia, unless it keeps rising and rising.

Also, energy prices have an unusual and dysfunctional feature compared with most of the rest of the economy. Whether it is turning on an electric light, setting a thermostat, or driving a car, we use the energy at one point in time and pay for it much later. This undermines both energy conservation generally and carbon pricing specifically. If we could see, in real time, how slowing down on the freeway or turning down our air-conditioning was affecting our payments, we would be more likely to change behavior. As the information revolution continues, this kind of real-time monitoring may become more common, and there are already some phone apps to report on energy use and cost in real time in some places.

Probably the main effects of carbon pricing will be not on individual energy conservation (turning down the thermostat) but rather on fuel choices at a larger scale. When an electric utility has to pay for carbon pollution from coal, and when nuclear power receives the same treatment as other low-carbon sources, then the economy can more effectively transition from fossil fuel to clean sources.

Figuring out the right level for any carbon pricing system is not simple, but we do know that it's not zero. The main thing is to put a fee in place and begin charging to dump CO_2 in the air. Sweden has tinkered with its price over the years, but overall it has worked.

Cap and Trade

An alternative way to price carbon, similar to a carbon tax/fee, is "cap and trade." In this system, a government sets a limit on how much carbon can be released (the cap) and issues permits allowing companies to do so. Sometimes the permits are free, and sometimes they are auctioned off by governments. Then companies can trade the permits, creating a market for CO_2 pollution in which efficiencies can be achieved.

These cap-and-trade systems exist around the world, but so far they have been less efficient than hoped, with prices fluctuating wildly. An economic downturn can suddenly reduce demand for permits and cause the price to crash. It's hard to get the price right, especially if governments start out giving away permits for free.

The European Union has an Emissions Trading System (ETS).[15] The system grants allowances to 11,000 heavy industrial and power installations as well as airlines, covering almost half of the EU's carbon emissions. It aims to cut CO_2 emissions by 20 percent, compared with 1990 levels, by 2020, and 40 percent by 2030. In the 2013–2020 period, emissions caps are being reduced by a bit less than 2 percent each year. The aviation sector has separate, less ambitious, targets. Allowances are initially given or sold to polluters by governments, and then the companies can trade them with each other as needed to match their pollution levels. A company with too high a level of emissions can either become more efficient or buy allowances from a less polluting

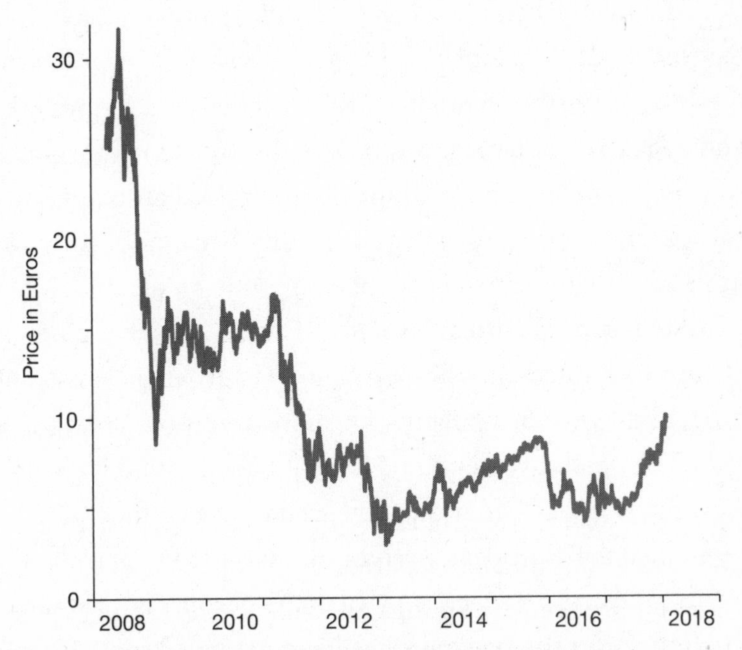

Figure 52. ETS emission-permit prices per ton CO_2 equivalent since 2008. *Data source*: European Environment Agency.

company. Of course, the company might instead shift production to another part of the world not covered by carbon pricing; this is an ongoing concern of the EU.

The EU's effort to price carbon pollution could play a positive role as the system matures. But to date it has had a bumpy start. In the first phase, 2005–2007, too many allowances were issued, and the price dropped to zero. In 2008–2012, the major economic recession reduced industrial activity, and hence emissions, and the price again dropped sharply. In 2013–2020, the system evolved, with fewer allowances given away for free and more being auctioned initially before being traded.[16] However, the price has remained low since 2013, below $10/ton—well below the low end of the estimates of the "social cost of carbon," and too low to significantly affect the trajectory of global warming. The European countries are indeed reducing their carbon emissions modestly, but mostly because of technology shifts, higher efficiency, and economic recessions rather than the cap-and-trade system. The EU is changing its system to reduce emissions permits faster when emissions drop, and Sweden is pushing for more ambitious changes in the ETS, which Sweden finds to be ineffective to date.[17]

California has had a cap-and-trade system since 2012. It resembles the European system in charging for permits for major industries. So far the price of the permits is only about $14/ton.[18] But the program has been extended to 2030 and might have more effect as the cap comes down. California is trying to integrate its system with others in the Pacific Northwest, especially in the western Canadian provinces, to create a bigger and more efficient trading market for permits.

A cap-and-trade system in the US Northeast, the Regional Greenhouse Gas Initiative, applies only to electricity generation, where it helps to integrate carbon-free sources onto the electric grid along with fossil sources. So far the carbon price is low, but the system has potential for the future, especially if applied more broadly beyond the electric sector, as California's system does.

China is just beginning an ambitious carbon trading program whose results are not yet known.[19] Despite strong support from the central government, the program will face challenges in data unreliability, corruption, and power struggles between local and national leadership. But if successfully implemented, the new program could be a positive force in fighting climate change. The initial program unveiled at the end of 2017 applies to the electric sector—about half of China's emissions—and has modest goals for emission reductions.[20]

The advantage of cap and trade is that an authority can set the actual allowable level of carbon pollution, whereas under a carbon tax, nobody can say for sure how many people will pay a given price and therefore how much CO_2 will go into the atmosphere.[21] But as the ETS and other efforts indicate, it is hard for governments to assess how many pollution permits to allow and how much to initially charge for them.

A carbon tax is simpler than an emissions trading system. It affects everything in a whole economy, is easy to administer, and acts quickly. As it restructures the economy away from fossil fuels, it can boost economic growth and create jobs (for example, wind-turbine installer or nuclear technician rather than gas-station attendant or coal miner). However, to achieve rapid decarbonization, a carbon price has to rise quickly and inexorably. So far, Sweden's use of a high carbon price is the exception in a world where politicians are skittish about raising taxes.

CHAPTER FIFTEEN

Act Globally

THE EXAMPLE OF Sweden shows that rapid decarbonization is possible. As one physicist put it recently, "Solving global warming does not require us to 'tear down capitalism.' The world just needs to be a bit more like Sweden."[1] Sweden's success has actually been multifaceted—resulting from a mix of plentiful hydropower, an early decision to reduce fuel imports, and a culture that values efficiency. But the main driver of success, shifting from fossil fuels to nuclear power, has been successfully implemented in other countries as well, notably France.

Although the 1970s and '80s saw Sweden, France, Belgium, Switzerland, and Finland quickly build out their nuclear power capacity, the model is not limited to Europe or to that time period. The Canadian province of Ontario shows the feasibility, in this century, of replacing coal with nuclear power.[2] With a population of 14 million, somewhat larger than Sweden's, Ontario is the industrial heartland of Canada. When it rolled out nuclear power plants, it built 16 reactors in seventeen years, 1976–1993.[3] Then in 2003–2014, it upgraded the province's nuclear power stations to bring nuclear power from 42 percent to 60 percent of the total (hydropower supplied most of the rest). In 2014 Ontario closed its last coal-operated power plant. In one decade, CO_2 emissions from Ontario's electric sector had fallen by almost 90 percent, with fossil fuels (all of it now methane) reduced to a small fraction.[4]

The model could be replicated around the world.[5] As we have seen, a country such as South Korea or Russia that pursues nuclear power expansion as a national policy can quickly produce nuclear reactors one after another, safely and economically. The director of the IAEA recently estimated the global investment needed to build 10–20 new reactors annually at $80 billion per year—more than doubling nuclear power capacity by 2040.[6] This is equivalent to one-tenth of 1 percent of the world's annual economic activity, is an investment not an expenditure, and is readily achievable given the political will. But actually, at $4–$8 billion per reactor, the price is probably higher than necessary, given that South Korea can build one for $2 billion per gigawatt today,[7] and the costs would come down in a period of repeated builds and economies of scale.

The idea of a large-scale world build-out of nuclear power as a response to climate change is still controversial but receiving new and serious attention from experts as the climate crisis gets worse.[8] Nuclear power risks are far smaller than climate change risks. In 2015 four leading climate scientists, who understand the climate risks, argued that "the only viable path forward" was an "accelerated deployment of new nuclear reactors" alongside the growth of renewables. Pointing to the examples of Sweden and France, they note that building 115 new reactors each year would entirely eliminate fossil fuels from the world's growing electricity production by 2050.[9] This scale of action, much more ambitious than that of the IAEA director, is the scale on which we should be thinking.

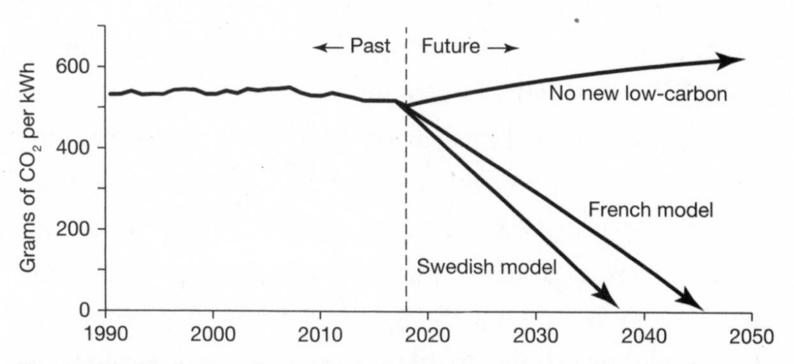

Figure 53. Carbon emissions intensity (per unit of electricity) worldwide, with no new low-carbon power and if low-carbon installations were built at the historical GDP-normalized rates of Sweden or France. *Data source*: International Energy Agency (emissions intensity and demand projection), OECD (GDP projection), and BP (electricity generation data).

Building 115 reactors per year may seem a large undertaking, but to put it in perspective, the Swedes had no problems building about 1 reactor per ten million inhabitants per year in the 1970s and early 1980s. That rate, applied globally today, would produce about 750 reactors each year, more than six times the proposed rate. At even half Sweden's rate, the world could eliminate fossil fuels from electricity generation by 2040 instead of 2050, *and* displace coal as a heat source for buildings and industrial use, *and* have electricity for energy-intensive applications such as producing fossil-free liquid fuels or sequestering CO_2 from the atmosphere. Many millions of lives would be saved, and billions of people would be economically uplifted. We know from cases like Sweden and France that this is doable, and we have much better technology and knowledge to do it now than they did several decades ago.

Because coal is the number-one problem for climate change, and coal generates round-the-clock power, an intensive program to build nuclear power plants worldwide is the most effective solution. Such a program cannot succeed haphazardly. It has to be a national decision, or countries will get stuck in the US-style swamp of too many designs and too little experience building them, with costs escalating.

Standardized design is a key aspect. In 1995 the head of the US NRC summed up the difference between the French and American nuclear industries thus: "The French have two kinds of reactors and hundreds of kinds of cheese, whereas in the United States the figures are reversed."[10] Other important elements for fast, low-cost construction are multiple builds of the same design to gain experience with it, siting multiple reactors at one location, strong government support, and centralized policy making through a single utility or agency.

China is a key player. It could pick a limited number of standardized designs and concentrate on a large-scale short-term expansion to phase out coal. (Stopping Chinese coal burning is the immediate top priority to save the planet.) At the same time, China could keep experimenting with a few fourth-generation designs to find out what will best follow in a decade. Replacing China's coal plants with nuclear power plants, at a Sweden-style pace, could be the single most important action anyone could take to combat climate change worldwide.

In the countries of the West, those with existing nuclear reactors could immediately stop shutting them down ahead of their useful life. Although limited in scale, this would be the cheapest, quickest way to get CO_2-free electricity. Countries where building more of the same appears hopeless from a political perspective, such as the United States and Germany, could get behind a few fourth-generation designs and build them out rapidly. The first nuclear reactors in history, such as Admiral Rickover's, were designed and built in a few years. Today we have more money, much better technology, and a better-educated workforce, so we can do at least as well.

Countries that can still build practical, affordable third-generation reactors, such as South Korea, could build out the fleet to power the country and get off fossil fuels, while exporting to other countries (as South Korea has done successfully in the UAE).

Russia could also step up its construction of new plants, continue to push its exports, and move forward aggressively with its fourth-generation Breakthrough technology development.

Decarbonization requires moving the world's electricity generation system rapidly from a carbon-intensive model, currently over 500 grams of CO_2 per kWh produced, to one with a tenth that level of carbon pollution per kWh. France,

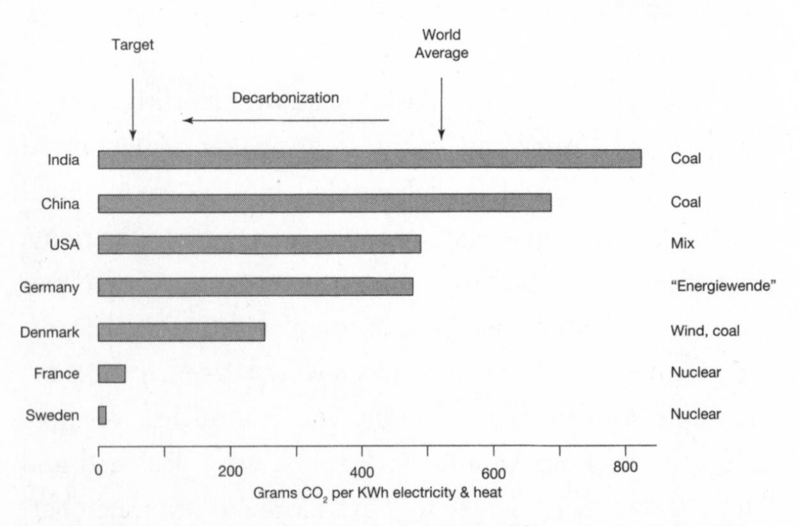

Figure 54. Carbon intensity of electricity in different countries. *Data source*: International Energy Agency data set (emissions per kWh of electricity and heat output).

Sweden, and Ontario have done just that; others are nowhere near and need to catch up.

Revamping the Economy

Beyond replacing coal and methane on the electric grid, we need to decarbonize across the economy. Of course, we want economic growth to continue, particularly in the poorer countries, but decoupled from fossil-fuel use.[11] As a very rough rule, energy today divides into three main parts—electricity, transportation, and heat. One simple formula for cleaning up the economy and getting rid of fossil

Figure 55. Electrify everything: Siemens is testing an electric highway for heavy trucks in Sweden. *Photo*: Courtesy of Siemens. www.siemens.com/press.

fuels is this: (1) generate electricity cleanly; (2) electrify everything.[12]

This transition is under way already. Electric trains have replaced diesel trains in many countries. Hybrid and all-electric cars are growing, and several countries have declared a goal to eliminate gasoline and diesel cars. Around the world, research on batteries and other aspects of electric cars can accelerate the transition from fossil to electric vehicles. The steel industry (in Sweden) has invented a new method to use electricity rather than coal in making steel.[13]

Sweden has also switched over to a combination of electric heating, notably with heat pumps, and "district heating" to heat multiple buildings at once, more efficiently and cleanly (biomass plays an important part in district heating). Perhaps the American suburbs will never be able to use district heating, but cities can, and even in the suburbs electric heat pumps are already growing in popularity and coming down in cost.

While electrification will undoubtedly be a main route to decarbonization, there are also great opportunities to directly supply cleanly produced heat. Conventional nuclear reactors are very well suited to supply heat to district heating networks, something that is already happening at fifty-seven reactors around the world, mainly in Russia. China has just finished construction of a novel high-temperature gas-cooled reactor[14] that can produce steam at 567°C, far higher than in conventional light-water reactors. The very high temperatures mean that electricity production is highly efficient (42 percent), and they also allow high-temperature industrial process heat supply, directly displacing coal plants at industrial facilities.

Another alternative to straightforward electrification is to create liquid and gaseous fuels to replace fossil fuels. This could be especially useful if cheap batteries prove elusive. Nuclear electricity could split water into hydrogen and oxygen, and that hydrogen gas could replace methane (with modifications to existing infrastructure).[15] Hydrogen is just methane without the carbon. Nuclear power could also be transformed into ammonia as an energy carrier, since ammonia contains only hydrogen and nitrogen, readily obtainable from water and air if you have energy.[16] Ammonia is energy denser than hydrogen and can be stored at lower pressures. Using nuclear power to create liquid or gaseous fuels has the advantage that the round-the-clock power of the reactor could be used for energy on the grid when demand is high and for fuel production when grid demand is low. This could also help integrate renewables onto the grid, as nuclear power could work around their ups and downs efficiently, producing liquid fuels when renewable production is high and grid electricity when renewables drop out.

Finally, serious carbon pricing would help revamp the economy away from fossil fuels. International agreements to standardize carbon prices across borders would make this easier to administer, provided the price was high enough to bring about the needed changes.

Politics

Ultimately, replacing fossil fuels with nuclear power is a political as much as a technical issue.[17] Politics does matter; if nuclear power is too problematic politically, it won't be able to make a difference. Even in Sweden, politics is threatening the success of nuclear power. In 2016 a government that included the Green Party wanted to phase out nuclear power, although in the end a compromise deal was reached that limits the damage to the Swedish electricity supply and the climate. However, current policies are still driving four well-operating Swedish reactors into very early retirement in the next few years.[18] A premature shutdown of the country's full nuclear power capacity would take clean energy off of a northern European grid reliant on German and Polish coal plants, leading to an estimated 50,000 energy-production-related deaths.[19]

Thinking about a potential major expansion of nuclear power worldwide, is politics really such a big constraint? China, the most central player for rapid decarbonization, is a one-party state where antinuclear organizations have little leverage. Russia, the key player in both current exports and fourth-generation technology, is also authoritarian and supportive of nuclear power.

If countries such as Germany and Japan want to shut off nuclear power and burn methane and coal instead, this is certainly a step backward for climate efforts but hardly a death blow for the prospects of a nuclear power build-out.

The United States is perhaps the most problematic case because it is the original source of nuclear expertise and today generates more total nuclear power than any other country, with twice the production of second-place France.[20] The lagging state of the US nuclear power industry is much more of a concern than the problems in other Western democracies. American politics has challenged the nuclear power industry for decades, with antinuclear groups being large, well-funded, and very active in lobbying, shaping public opinion, and litigation. As we have seen, in a supreme irony, the very groups most actively opposing nuclear power are those most vocal about climate change. On the hopeful side, there is a growing pronuclear wing of the environmentalist movement[21] and literature[22] in the United States, although it is still quite small, entirely reliant on private donations, and has access to just a tiny fraction of the financial resources of the antinuclear movement.

American politics are not, however, altogether unfavorable to a nuclear power expansion. States have begun to realize that existing nuclear power plants play a key role in meeting emissions targets and have begun in a few cases to treat nuclear power on a more level playing field with renewables. This trend could lead to a shift from Renewable Portfolio Standards—state mandates to include a certain percentage of renewables in the energy mix—to Low-Carbon Portfolio Standards that include nuclear power along with renewables.[23]

There is also some bipartisan consensus regarding clean energy, which receives considerable support even from the right of the American political spectrum. Archconservative governor of Kansas Sam Brownback was a big supporter of the large-scale expansion of wind power in that state. It was good for the Kansas economy. Other top wind-power states are also mostly Republican leaning.[24] And concerns about climate change, although taboo among Republican elected officials, are actually strong among rank-and-file Republican voters, with about half saying they would be more likely to vote for a candidate who supports fighting climate change.[25]

Nuclear power, especially for fourth-generation designs, has considerable bipartisan support in the US Congress. Liberals are becoming more open to nuclear power because of climate change, and conservatives support it for such reasons as economics and US superiority in technology.[26]

The antinuclear movement has progressed through reasons to oppose nuclear power, one after another. First, it was that nuclear power was too dangerous, but after fifty years with an exemplary safety record, this argument gets thin. Next, nuclear power would lead to the proliferation of nuclear weapons, including to terrorists. Again, the record shows otherwise.[27]

Then it was that nuclear power was simply uneconomical because US and European efforts to build new plants came in way over budget and behind schedule. These economic issues, which we address shortly, were partly the

Figure 56. German antinuclear protesters, 2011; no such enthusiasm to stop coal. *Photo*: Rot-Braun Magdeburg via Wikimedia Commons (CC-BY-2.0).

result of the constant litigation and regulatory demands of the antinuclear groups themselves. Nuclear power in Sweden has been highly competitive for decades, and in places like South Korea newer plants are similarly affordable.

The next antinuclear argument was that "we don't need nuclear power" because renewables will solve everything. But in every case where nuclear power was shut down, renewables have not filled the gap and CO_2 emissions have gone up,[28] whereas in places such as Ontario that expanded nuclear power, emissions went down.

In the end, those who have spent decades making nuclear power politically unpopular argue that a nuclear power expansion is simply politically infeasible. This is just a self-fulfilling prophesy, and we should not be so quick to write off the most practical solution for humanity's most serious problem as politically undoable. Politics have a way of catching up with necessity.

The political problems that appear most challenging are specific not to nuclear power but to tackling climate change itself. As we have seen, the international community's best effort to date, the 2015 Paris Agreement, does not come very close to solving the problem and, even for its inadequate goals, does not have any enforcement mechanism for each country's voluntary commitments. In fact, many countries are falling short of their Paris commitments, yet the international political system does not have a mechanism to correct this, much less to push them further in the right direction.[29]

Think again about the idea of a large asteroid heading for a collision with Earth. Would we ask countries for voluntary contributions and then let them fall short without consequence? No, we would create international structures and processes to reliably solve the problem, drawing on all the world's available talent and resources.

Economics

The argument that nuclear power these days is "too expensive" ignores the places where nuclear power has been cost competitive, such as in South Korea, Russia, and of course Sweden. It also ignores cases such as Vermont Yankee, where nuclear power plants have closed long before their licensed end dates, despite producing energy more cheaply than any source other than methane. Environmentalists who oppose nuclear power demand short-term profitability without subsidies but make no such demand on renewables. Nuclear power plants require large up-front investments but then produce large amounts of clean power for sixty years or more, several times the life span of wind or solar installations.[30] Nuclear power does not require expensive storage because it is not intermittent. Without a carbon price, nuclear power does not compete with cheap German coal or US methane, but it can be cheaper than most renewables without their subsidies.[31] And in South Korea in 2013, nuclear power generated electricity at 3.7 cents/kWh, cheaper than coal (5.6 cents), hydro (16.2 cents), or LNG (20.5 cents).[32] "Too expensive" clearly does not justify environmentalist demands to shut down South Korea's nuclear power plants.

Large capital expenditures at the start of a long-term project such as a nuclear power plant do not come easily in a deregulated private marketplace, especially when prices of competing fossil fuels are unstable. Also, the discount rate, discussed above in regard to carbon pricing, strongly affects the economics of a long-term investment such as a nuclear power plant. One analysis of the cost of energy in Western industrialized countries, "levelized" over the whole life cycle of the technology from mining to decommissioning, found that at a 3 percent discount rate, nuclear power came in cheaper than coal or methane. But at a 10 percent discount rate, it lost that advantage.[33]

In some ways building a nuclear power plant is more like building a railroad than a gas station, and long-term government support has been important to ensure the success of such large-scale long-lived projects in an unstable market.[34] South Korea generates cheap electricity because, with repeated builds, it can construct a nuclear power plant for about

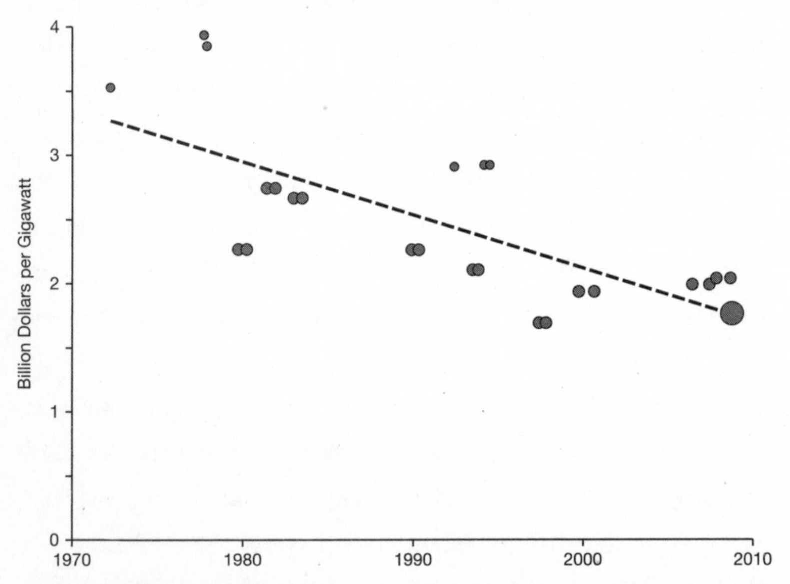

Figure 57. Nuclear power plant construction costs in South Korea, 1972–2008. Dot sizes indicate capacities of 558–669 MW, 903–1,001 MW, and 1,300+ MW. *Data source*: Courtesy of Jessica Lovering based on Jessica R. Lovering, Arthur Yip, and Ted Nordhaus, "Historical Construction Costs of Global Nuclear Power Reactors," *Energy Policy* 91 (April 2016): 371–382.

$2 billion per gigawatt. In the United Kingdom, it now costs $8 billion per gigawatt, and in the United States the latest attempt runs to around $12 billion.[35] This does not mean nuclear power is "not economical" but rather that Britain and the United States should start to do what South Korea did—build successive plants of the same design with government backing for the long-term investment. A recent study of cost drivers in nuclear plant construction around the world emphasizes efforts to capture learning during the buildout of a fleet and completion of designs prior to the start of construction.[36]

Beyond Costs and Burdens

To date, the international politics of climate change have been framed far too much around a narrative of burden sharing. We assume that the solution to a future disaster will be expensive to the present generation and that the international community's big role is to allocate those costs among the world's sovereign countries. This approach guarantees that everyone wants someone else to pay the costs and that today's inhabitants of Earth will want to put off costs to an indeterminate future.

How could this narrative be reframed away from costs and burdens and toward opportunities and inventions? Swedes are not moping around suffering from burdens because they cut their CO_2 emissions. On the contrary, their economy is robust, they are healthier with cleaner air, and they use plenty of energy to stay warm and well lighted in the Scandinavian winter. Similarly, people who drive electric cars do not suffer because the cars perform badly but, at least, pollute less. Not at all. Electric cars are fun to drive, and hundreds of thousands of people have lined up to buy Tesla Model 3s before the first one was even produced—not because we all want to suffer to save the world, but because Teslas are cool. With nuclear power and renewables together, we have the opportunity to make energy available to poor people around the world, which should lead to better health, less armed conflict, and lower population growth. Those are good things.

Figure 58. Electric cars and other energy innovations are fun, not burdensome. *Photo*: Courtesy of Steve Cole.

Acknowledgments

From Joshua Goldstein: For education and discussion of climate change and nuclear power issues, I thank Matt Wald (Nuclear Energy Institute); Robert Orr, Christopher Foreman, and the late John Steinbrunner (University of Maryland); Michael Oppenheimer, Robert Keohane, M. V. Ramana, Harold Feiveson, Frank von Hippel, and Rob Socolow along with the participants in his seminar (Princeton); Josh Freed (Third Way); Allison Macfarlane (George Washington University); Aaron Bernstein (Harvard Medical); Rachel Pritzker (Pritzker Innovation Fund); Ted Nordhaus and Peter Teague (Breakthrough Institute); Michael Shellenberger (Environmental Progress); Armond Cohen and Ashley Finan (Clean Air Task Force, and extra thanks to Armond for help spreading the message); Krister Svahn, Mats Ladeborn, Oskar Ahnfelt, and Johan Pettersson (Vattenfall/Ringhals); Martin Cohn (Vermont Yankee); Robert Hargraves (ThorCon); Eric Ingersoll; Jacopo Buongiorno (MIT); Meredith Angwin; Representative Solomon Goldstein-Rose (MA); Michael Lynch (Energy SEER); Peter Haas (University of Massachusetts); and Lisa Martin, Helen Kinsella, Ron Mitchell, Kate Ricke, and other participants in the 2017 "Social and Political Dimensions of Climate Change" workshop at the University of Wisconsin, Madison. For access to research resources, thanks to the University of Massachusetts Library and Political Science Department. For help with ideas and writing, thanks to Steven Pinker, Fredrica Friedman, Jon Pevehouse, Andra Rose, Michael Goldstein, and Fred Goldstein-Rose. Finally, my partner, J. Alden Cox, has played an indispensable role in the development of the book.

From Staffan Qvist: For helpful suggestions, thanks to Malwina Qvist, Björn and Ninni Qvist, Robert Stone, and James Hansen. Special thanks to Daniel Westlén for extensive comments on a draft.

From both: For help using their data, illustrations, and photographs, we thank Jessica Lovering, Nickolay Lamm, Kate Lamb, Hanno Böck, Ibrahim Malik, Terese Winslow, Steve Cole, Climate Interactive (Ellie Johnston), Environmental Defense Fund (Kelsey Robinson), World Resources Institute (Maria Hart), Nuclear Threat Initiative (Meaghan Webster), Vattenfall (Krister Svahn), SKB (Simon Roth), ČEZ (Martin Schreur), NuScale Power (James Mellott), Siemens (Oliver Schmitt), and Getty Images (Samuel), as well as Rebecca Martin, Margarita Diaz, and Sewela Mamphiswana for images that did not end up being used.

Our literary agent, Max Brockman, expertly directed the book to the right publisher, and our editor at PublicAffairs, Ben Adams, deftly guided the process from there. Thanks also to production editors Sandra Beris and Michelle Welsh-Horst, copyeditor Annette Wenda, and the rest of the PublicAffairs team and to our independent publicist, Leah Paulos. Finally, thanks to Klara Ingersoll for developing graphics and related materials to promote the ideas in this book.

Notes

1. Climate Won't Wait

1. We use the terms *global warming* and *climate change* interchangeably. Global warming produces climate change. Opinion polling shows the public does not differentiate the terms. See Riley E. Dunlap, "Global Warming or Climate Change: Is There a Difference?," *Gallup News*, April 22, 2014, http://news.gallup.com/poll/168617/global-warming-climate-change-difference.aspx.
2. J. G. J. Olivier, K. M. Schure, and J. A. H. W. Peters, *Trends in Global CO_2 and Total Greenhouse Gas Emissions: 2017 Report* (The Hague: PBL Netherlands Environmental Assessment Agency, December 2017).
3. Joeri Rogelj et al., "Paris Agreement Climate Proposals Need a Boost to Keep Warming Well Below 2°C," *Nature* 534 (June 30, 2016): 631–639; Glen P. Peters et al., "Key Indicators to Track Current Progress and Future Ambition of the Paris Agreement," *Nature Climate Change* 7 (2017): 118–122.
4. World Bank data. See Figure 18 in Chapter 5.
5. BP, *BP Statistical Review of World Energy 2017*, 9.
6. Intergovernmental Panel on Climate Change (IPCC), *Climate Change 2014: Synthesis Report. Contribution of Working Groups I, II and III to the Fifth Assessment Report of the Intergovernmental Panel on Climate Change* [core writing team, R. K. Pachauri and L. A. Meyer, eds.] (Geneva: IPCC, 2014).
7. Johan Rockström et al., "A Roadmap for Rapid Decarbonization," *Science* 355, no. 6331 (2017): 1269–1271.

8. National Oceanic and Atmospheric Administration, "2016 Marks Three Consecutive Years of Record Warmth for the Globe," January 18, 2017, www.noaa.gov/stories/2016-marks-three-consecutive-years-of-record-warmth-for-globe.
9. Climate Central, "Extreme Sea Level Rise and the Stakes for America," April 26, 2017, www.climatecentral.org/news/extreme-sea-level-rise-stakes-for-america-21387.
10. Derek Watkins, "China's Coastal Cities, Underwater," *New York Times*, December 11, 2015.
11. National Snow and Ice Data Center, "Quick Facts on Ice Sheets," https://nsidc.org/cryosphere/quickfacts/icesheets.html.
12. John Vidal, "'Extraordinarily Hot' Arctic Temperatures Alarm Scientists," *Guardian*, November 22, 2016.
13. D. A. Smeed et al., "Observed Decline of the Atlantic Meridional Overturning Circulation, 2004–2012," *Ocean Science* 10 (2014): 29–38; Quirin Schiermeier, "Atlantic Current Strength Declines," *Nature* 509 (2014): 270–271; Robinson Meyer, "The Atlantic and an Actual Debate in Climate Science," *Atlantic*, January 7, 2017.
14. Angela Fritz, "Boston Clinches Snowiest Season on Record amid Winter of Superlatives," *Washington Post*, March 15, 2015.
15. David Wallace-Wells, "The Uninhabitable Earth," *New York*, July 10, 2017.
16. Michael L. Klare, *The Race for What's Left: The Global Scramble for the World's Last Natural Resources* (London: Picador, 2012); Shlomi Dinar, ed., *Beyond Resource Wars: Scarcity, Environmental Degradation, and International Cooperation* (Cambridge, MA: MIT Press, 2011).
17. See, for example, https://climateandsecurity.org/; and Joshua S. Goldstein, "Climate Change as Global Security Issue," *Journal of Global Security Studies* 1, no. 1 (2016).
18. Steven Pinker, *The Better Angels of Our Nature: Why Violence Has Declined* (New York: Viking, 2011); Joshua S. Goldstein, *Winning the War on War: The Decline of Armed Conflict Worldwide* (New York: Dutton, 2011).

19. Solomon M. Hsiang, Marshall Burke, and Edward Miguel, "Quantifying the Influence of Climate on Human Conflict," *Science* 341, no. 6151 (2015): 1212.
20. Bhadra Sharma and Ellen Barry, "Quake Prods Nepal Parties to Make Constitutional Deal," *New York Times*, June 9, 2015: A6; Andrew M. Linke et al., "Rainfall Variability and Violence in Rural Kenya," *Global Environmental Change* 34 (2015): 35–47.
21. Jan Selby et al., "Climate Change and the Syrian Civil War Revisited," *Political Geography* 60 (September 2017): 232–244.
22. Adrian E. Raftery et al., "Less than 2°C Warming by 2100 Unlikely," *Nature Climate Change* 7 (2017): 637–641.
23. Intergovernmental Panel on Climate Change, *Global Warming of 1.5°C*, Special Report, October 6, 2018; James Hansen et al., "Ice Melt, Sea Level Rise and Superstorms: Evidence from Paleoclimate Data, Climate Modeling, and Modern Observations That 2°C Global Warming Could Be Dangerous," *Atmospheric Chemistry and Physics* 16 (2016): 3761–3812.
24. James Hansen et al., "Young People's Burden: Requirements of Negative CO_2 Emissions," *Earth System Dynamics* 8 (2017): 577–616; James Hansen et al., "Assessing 'Dangerous Climate Change': Required Reduction of Carbon Emissions to Protect Young People, Future Generations and Nature," *PLoS ONE* 8, no. 12 (2013): e81648.
25. *Juliana v. United States*. See ourchildrenstrust.org.
26. Naomi Klein, *This Changes Everything: Capitalism vs. the Climate* (New York: Simon & Schuster, 2014), 10.
27. George Marshall, *Don't Even Think About It: Why Our Brains Are Wired to Ignore Climate Change* (New York: Bloomsbury, 2014).
28. While trained and active as a nuclear engineer, Qvist currently leads a solar energy initiative in East Africa. Goldstein has solar panels on his roof.
29. Christiana Figueres et al., "Three Years to Safeguard Our Climate," *Nature* 546 (June 28, 2017): 593–595.
30. C-ROADS model at www.climateinteractive.org/tools/c-roads/ or the *New York Times* version at www.nytimes.com/interactive/

2017/08/29/opinion/climate-change-carbon-budget.html. Scientific review of the model is at www.climateinteractive.org/wp-content/uploads/2014/01/C-ROADS-Scientific-Review-Summary1.pdf.

31. This is a simple linear reduction from the baseline value in 2020 rather than a running reduction.
32. This calculation is based on the 2°C scenarios in Glen P. Peters et al., "Key Indicators to Track Current Progress and Future Ambition of the Paris Agreement," *Nature Climate Change* 7, no. 2 (2017): 118–122.
33. S. Pacala and R. Socolow, "Stabilization Wedges: Solving the Climate Problem for the Next 50 Years with Current Technologies," *Science* 305, no. 5686 (2004): 968–972; Steven J. Davis et al., "Rethinking Wedges," *Environmental Research Letters* 8 (2013): 011001.
34. Peter J. Loftus, "A Critical Review of Global Carbonization Scenarios: What Do They Tell Us About Feasibility?," *WIREs Climate Change* (2013), doi:10.1002/wcc.324.
35. In addition to converting to electric heating, industrial facilities and district heating networks can use heat created in the process of generating electricity in low-carbon-emissions thermal power plants that use steam to generate electricity.
36. US Energy Information Administration, *Annual Energy Outlook, 2018* (February 6, 2018), 81, www.eia.gov/outlooks/aeo/pdf/AEO2018.pdf.

2. What Sweden Did

1. Staffan A. Qvist and Barry W. Brook, "Potential for Worldwide Displacement of Fossil-Fuel Electricity by Nuclear Energy in Three Decades Based on Extrapolation of Regional Deployment Data," *PLoS ONE* 10, no. 5 (2015): e0124074.
2. The first outright ban from the Swedish government on hydroelectric power expansion passed in April 1970, effectively ending the period of expansion that lasted from the late 1800s.

3. Gwyneth Cravens, *Power to Save the World: The Truth About Nuclear Energy* (New York: Vintage, 2007), 60.
4. A typical reactor producing 6 TWh per year is equivalent to burning 2.75 million tons of coal, with a railcar holding 100–120 tons.
5. Bruno Comby, *Environmentalists for Nuclear Energy* (1994; English translation, Paris: TNR Editions, 2001), 45.
6. A recent review estimates nuclear power's "total spatial footprint" (including uranium mining) at about 1/13th that of coal power (and, at most 1/4 that of wind, 1/50th that of solar photovoltaics and 1/1500th that of biomass). Vincent K. M. Cheng and Geoffrey P. Hammond, "Energy Density and Spatial Footprints of Various Electrical Power Systems," *Energy Procedia* 61 (2014): 578–581.
7. Sven Werner, "District Heating and Cooling in Sweden," *Energy* 126 (2017): 419–429. In the 1960s, most Swedish buildings used fuel oil for heat; district heating accounted for only 3 percent of the heat market in 1960. Today, district heating accounts for 58 percent of the energy for heating buildings (2014), while fuel oil accounts for less than 2 percent. The remaining heat sources consist of electricity used for electric heating and heat pumps and a very small amount of methane.
8. Qvist and Brook, "Potential for Worldwide Displacement of Fossil-Fuel Electricity." See note 1.
9. In 2016, with shutdowns for maintenance and upgrades, Sweden's plants produced electricity at an average 75 percent of capacity.
10. https://corporate.vattenfall.se/om-oss/var-verksamhet/var-elproduktion/ringhals/ringhals-nuclear-power-plant/.
11. See comparisons in Comby, *Environmentalists for Nuclear Energy*, 61–62.
12. Mara Hvistendahl, "Coal Ash Is More Radioactive Than Nuclear Waste," *Scientific American* (December 13, 2007).
13. Calculated from Staffan A. Qvist and Barry W. Brook, "Environmental and Health Impacts of a Policy to Phase Out Nuclear Power in Sweden," *Energy Policy* 84 (2015): 1–10.

14. Calculated using emissions intensity factors from the IPCC Contribution of Working Group III to the Fifth Assessment Report of the Intergovernmental Panel on Climate Change, and the average Ringhals Power Plant production for 1999–2016 of 24.1 TWh per year.
15. The operation of kärnkraft produces no emissions at all (just like wind, solar, or hydroelectric power), but all sources of energy have associated "life-cycle" emissions, taking into account the emissions of, for instance, mining for the materials that make up the energy production unit. According to Swedish state utility Vattenfall, kärnkraft has the lowest total life-cycle greenhouse gas emissions of any known energy source.
16. The "capacity factor" of wind farms varies with technology, location, and type (offshore or onshore), between a low of around 14 percent (for poorly located onshore wind farms) up to 55 percent for the best-performing offshore wind farms. The production-averaged global wind capacity factor is around 33 percent.
17. Countries that have abundant hydroelectric power reservoirs can relatively cost-effectively "store" intermittent energy by simply running the hydro power less and saving water for later. Such opportunities, unfortunately, exist in only a few lucky countries.
18. Estimates about triple this size are in Nuclear Energy Institute, "Land Requirements for Carbon-Free Technologies," June 2015, www.nei.org/CorporateSite/mediafilefolder/Policy/Papers/Land_Use_Carbon_Free_Technologies.pdf.
19. OECD, "Radioactive Waste Management and Decommissioning in Sweden, [2012], 10–12, www.oecd-nea.org/rwm/profiles/Sweden_report_web.pdf. Twelve thousand tonnes of spent fuel, over fifty years from all Swedish reactors; 160,000 cubic meters of all waste, over fifty years. Current rate is 1,000–1,500 cubic meters per year. Spent fuel receiving capacity is 300 cubic meters per year. Ringhals accounts for 44 percent of the national total.
20. Stewart Brand, *Whole Earth Discipline: An Ecopragmatist Manifesto* (New York: Viking, 2009), 111.

21. While operational emissions are nearly zero, kärnkraft units do emit tiny amounts of CO_2 during ancillary operations such as testing backup diesel generators.

3. What Germany Did

1. Clean Energy Wire, "Germany's Energy Consumption and Power Mix in Charts." Data from AG Energiebilanzen 2017.
2. Melissa Eddy, "Missing Its Own Goals, Germany Renews Effort to Cut Carbon Emissions," *New York Times*, December 4, 2014, A6.
3. Vattenfall [utility company], "Energy from Lusatia: Jänschwalde Lignite Fired Power Plant," fact sheet, www.leag.de/fileadmin/user_upload/pdf-en/fb_kw_jaewa_10seiter_engl_2013.pdf. In an average year, Ringhals's electricity output is about 20 percent higher than that of Jänschwalde.
4. Ibid. Capacity is 82,000 tons daily, but the plant does not operate at capacity all the time. Based on electrical output and heat content of lignite, we estimated about 50,000 tons on average. In addition, large amounts of coal are burned to dry out the lignite and to operate the mining equipment.
5. Fifty thousand tons of lignite times 2,792 pounds of CO_2/ton based on US Energy Information Administration, "Carbon Dioxide Emissions Coefficients," February 2, 2016.
6. From Anil Markandya and Paul Wilkinson, "Electricity Generation and Health," *Lancet* 370 (2007): 981. Their estimate of 32.6 annual deaths and 298 serious illnesses per TWh for European lignite is multiplied by Jänschwalde's 20 TWh of production.
7. World Wildlife Federation, "Dirty Thirty: Ranking of the Most Polluting Power Stations in Europe," May 2017, http://d2ouvy59p0dg6k.cloudfront.net/downloads/european_dirty_thirty_may_2007.pdf.
8. Clean Energy Wire, "State Secretary Baake—Last German Lignite Plant Likely to Be Switched Off Between 2040 and 2045," News Digest Item, October 20, 2016.

9. LEAG, "The Lignite Power Plants," www.leag.de/en/business-fields/power-plants/.
10. Hubertus Altmann in Vattenfall, "Flexible and Indispensible: Lignite-Based Power Generation in the Energiewende," 2015, www.leag.de/fileadmin/user_upload/pdf-en/brosch_flexGen_en_final.pdf, 27.
11. Power Engineering, "Best Solar Project: GP Joule and Saferay's Solarpark Meuro in Germany," www.power-eng.com/articles/slideshow/2013/november/2012-projects-of-the-year/pg001.html.
12. Power Technology, "Fantanele-Cogealac Wind Farm," www.power-technology.com/projects/-fantanele-cogealac-wind-farm/.
13. Production in 2013 was about 1,250 gigawatt-hours.
14. Average wind capacity factors vary widely from region to region and are generally increasing with new and more efficient technology, new sites in very windy regions, and an increasing fraction of offshore wind. Individual offshore wind farms in ideal locations are able to reach capacity factors as high as 50 percent (Anholt-1 in Denmark), while the global average wind capacity factor today is about 23 percent (using generation data from BP Statistical Review and capacity data from Global Wind Energy Council, both for 2016). GWEC uses 30 percent capacity factor as a future, post-2030, average. See GWEC, "Global Wind Energy Outlook, 2016." Thirteen 600 MW wind farms operating at 30 percent capacity factor could potentially supply the same total electricity over a year as Jänschwalde (20 TWh/year).
15. Stanley Reed, "Power Prices Go Negative in Germany, a Positive for Consumers," *New York Times*, December 26, 2017, B3.
16. Calculated from national data on renewables production, mostly from grid operators.

4. More Energy, Not Less

1. Ranked ninth in World Bank 2015 data, energy use per capita.
2. The perceived relative "energy efficiency" of certain developed nations may also be deceptive. Britain uses comparatively

little energy per GDP, but this is mainly because a large fraction of its economy today is based on services rather than industry. Since British consumption of goods is not decreasing, it has simply outsourced its industrial production and, along with it, parts of its energy consumption and emissions, to China and other countries.

3. Steven Pinker, *Enlightenment Now: The Case for Reason, Science, Humanism, and Progress* (New York: Viking, 2018), 139–142; Charles C. Mann, *The Wizard and the Prophet: Two Remarkable Scientists and Their Dueling Visions to Shape Tomorrow's World* (New York: Alfred A. Knopf, 2018), 339–347.
4. *Individual* lifestyle changes of high-income environmentally conscious people unfortunately have relatively low impact on overall emissions. Research shows that "individuals with high pro-environmental self-identity intend to behave in an ecologically responsible way, but they typically emphasize actions that have relatively small ecological benefits." For instance, a detailed study of one thousand representative Germans finds that "energy use and carbon footprints were slightly higher among self-identified greenies." The primary determinant of a person's actual ecological footprint is income, followed by geography (rural versus urban), socioeconomic indicators (age, education level), and household size. The variables that most predict carbon footprint are "per capita living space, energy used for household appliances, meat consumption (so going vegetarian does indeed have real climate impact!), car use, and vacation travel." From Stephanie Moser and Silke Kleinhückelkotten, "Good Intents, but Low Impacts: Diverging Importance of Motivational and Socioeconomic Determinants Explaining Pro-environmental Behavior, Energy Use, and Carbon Footprint," *Environment and Behavior* (June 9, 2017).
5. David J. C. MacKay, *Sustainable Energy—Without the Hot Air* (Cambridge: UIT Cambridge, 2008), 68.
6. Kenneth Gillingham, David Rapson, and Gernot Wagner, "The Rebound Effect and Energy Efficiency Policy," *Review of Environmental Economics and Policy* 10, no. 1 (2016): 68–88.

7. Matt Piotrowski, "U.S. Shatters Record in Gasoline Consumption," *The Fuse*, February 28, 2017, http://energyfuse.org/u-s-shatters-record-gasoline-consumption/.
8. Shashank Bengali, "One Appliance Could Determine Whether India, and the World, Meet Climate Change Targets," *Los Angeles Times*, December 29, 2017.
9. International Energy Agency, *The Future of Cooling: Opportunities for Energy-Efficient Air Conditioning* (Paris: OECD/IEA, 2018), 59.
10. U.S. Energy Information Administration, "EIA Projects 48% Increase in Energy Consumption by 2040," May 12, 2016, www.eia.gov/todayinenergy/detail.php?id=26212.
11. The World Bank estimate for 2014 is 86 million (based on household surveys), while the International Energy Agency differs somewhat (based on utility connections). See World Bank, *Global Tracking Framework: Progress Toward Sustainable Energy, 2017*, annex 2.1, www.worldbank.org/en/topic/energy/publication/global-tracking-framework-2017, 37; International Energy Agency, *World Energy Outlook, 2016* (Paris: OECD/IEA, 2016), 92. Population growth is currently about 83 million per year.
12. US Energy Information Administration, *International Energy Outlook, 2016* (Washington, DC: US Energy Information Administration, 2016), 81–82.
13. Gayathri Vaidyanathan, "Coal Trumps Solar in India," *Scientific American/ClimateWire*, October 19, 2015.
14. Government of India, National Institution for Transforming India (NITI Aayog), *India Three Year Action Agenda, 2017–18 to 2019–20* (August 2017), http://niti.gov.in/writereaddata/files/coop/India_ActionAgenda.pdf, 99.
15. John Asafu-Adjaye et al., "An Ecomodernist Manifesto," April 2015, www.ecomodernism.org.
16. IEA/OECD statistics for 2014, from World Bank database.
17. World Bank data.
18. Clyde Haberman, "The Unrealized Horrors of Population Explosion," *New York Times*, Retro Report (online), May 31, 2015; Mann, *Wizard and Prophet*, 165–200; Gregg Easterbrook, *It's Better*

than It Looks: Reasons for Optimism in an Age of Fear (New York: PublicAffairs, 2018), 3–11.

5. 100 Percent Renewables?

1. See, for example, https://environmentmassachusettscenter.org/programs/azc/100-renewable-energy.
2. David J. C. MacKay, *Sustainable Energy—Without the Hot Air* (Cambridge: UIT Cambridge, 2008).
3. Eduardo Porter, "Why Slashing Nuclear Power May Backfire," *New York Times*, November 8, 2017, B1; Frankfurt School–UNEP Collaborating Centre / Bloomberg New Energy Finance, *Global Trends in Renewable Energy Investment, 2018* (Frankfurt am Main: Frankfurt School of Finance & Management, 2018).
4. Junji Cao et al., "China-U.S. Cooperation to Advance Nuclear Power," *Science* 353, no. 6299 (2016): 548. A critique of this article—Amory B. Lovins et al., "Relative Deployment Rates of Renewable and Nuclear Power: A Cautionary Tale of Two Metrics," *Energy Research & Social Science* 38 (2018): 188–192—contains a factor-of-ten error in the growth rate of nuclear power that negates the critique's conclusion.
5. Rauli Partanen and Janne M. Korhonen, *Climate Gamble: Is Anti-nuclear Activism Endangering Our Future?*, 3rd ed. (n.p.: CreateSpace, 2017), 34–35 (from Finnish edition, Janne M. Korhonen and Rauli Partanen, *Uhkapeli Ilmastolla* [Communications Agency CRE8 Oy, 2015]).
6. One-quarter of the 130 new TWh per year referenced in the last chapter.
7. US Energy Information Administration, "Chinese Coal-Fired Electricity Generation Expected to Flatten as Mix Shifts to Renewables," September 27, 2017, www.eia.gov/todayinenergy/detail.php?id=33092.
8. "BP Statistical Review of World Energy," June 2017, www.bp.com/content/dam/bp/en/corporate/pdf/energy-economics/statistical-review-2017/bp-statistical-review-of-world-energy-2017-full-report.pdf. These numbers have been adjusted to

account for the fact that solar and wind produce electricity directly, not through the conversion of heat: "The primary energy…from renewable sources [has] been derived by calculating the equivalent amount of fossil fuel required to generate the same volume of electricity in a thermal power station, assuming a conversion efficiency of 38% (the average for OECD thermal power generation)." Without this adjustment, renewables' share would be even lower.

9. "Requiem for a River: Can One of the World's Great Waterways Survive Its Development?," *Economist* (2017), www.economist.com/news/essays/21689225-can-one-world-s-great-waterways-survive-its-development.
10. BBC News, "Laos Dam Collapse: Many Feared Dead as Floods Hit Villages," July 24, 2018, www.bbc.co.uk/news/world-asia-44935495.
11. European Wind Power Association and European Commission, "Wind Energy: The Facts," 2009, www.wind-energy-the-facts.org, 219; Erik Magnusson, "Lillgrund ger ägarna stora förluster," *Sydsvenskan,* January 23, 2017.
12. "Lazard's Levelized Cost of Energy Analysis—Version 11.0," November 2017, www.lazard.com/media/450337/lazard-levelized-cost-of-energy-version-110.pdf, 2–3.
13. Eva Topham and David McMillan, "Sustainable Decommissioning of an Offshore Wind Farm," *Renewable Energy* 102, no. B (2017): 470–480.
14. www.dongenergy.co.uk/news/press-releases/articles/dong-energy-awarded-contract-to-build-worlds-biggest-offshore-wind-farm; www.morayoffshore.com/moray-east/the-project/. More problematically, the lower wind price sparked demands to cancel Britain's new nuclear plant. See "Nuclear Plans 'Should Be Rethought After Fall in Offshore Windfarm Costs,'" *Guardian,* September 11, 2017. See also UK Government, Department of Energy and Climate Change, "Investing in Renewable Technologies: CfD Contract Terms and Strike Prices," December 2013, www.gov.uk/government/publications/investing-in-renewable

-technologies-cfd-contract-terms-and-strike-prices, 7; and Partanen and Korhonen, *Climate Gamble*, 88.

15. Mark Harrington, "Wind Farm's Long-Term Cost Will Be High for Power Projects," *Newsday*, February 19, 2017; Diane Cardwell, "Way Is Cleared for Largest U.S. Offshore Wind Farm," *New York Times*, January 26, 2017, B3.
16. Peter Fairley, "Why China's Wind Energy Underperforms," *IEEE Spectrum* (May 23, 2016).
17. www.electricitymap.org/?wind=false&solar=true.
18. Ivan Penn, "California Invested Heavily in Solar Power. Now There's So Much That Other States Are Sometimes Paid to Take It," *Los Angeles Times*, June 22, 2017.
19. Ivan Penn, "Solar Power to Be Required for New Homes in California," *New York Times*, May 10, 2018, B10.
20. Pilita Clark, "Renewables Overtake Coal as World's Largest Source of Power Capacity," *Financial Times*, October 25, 2016.
21. US Energy Information Agency, "Levelized Cost and Levelized Avoided Cost of New Generation Resources in the Annual Energy Outlook, 2017, April 2017, www.eia.gov/outlooks/aeo/pdf/electricity_generation.pdf, 7.
22. Peter Maloney, "How Can Tucson Get Solar + Storage for 4.5¢/kWh?," *Utility Dive*, May 30, 2017, www.utilitydive.com/news/how-can-tucson-electric-get-solar-storage-for-45kwh/443715/.
23. "Lazard's Levelized Cost of Energy Analysis," 2–3. See footnote 12.
24. Varun Sivaram, *Taming the Sun: Innovations to Harness Solar Energy and Power the Planet* (Cambridge, MA: MIT Press, 2018).
25. Ibid., 73.
26. Electricitymap.org/.
27. Sivaram, *Taming the Sun*, 56–57, 78.
28. Ibid., 76.
29. Ibid.
30. Ibid., 64; Michael Shellenberg; "If Solar and Wind Are So Cheap, Why Are They Making Electricity So Expensive?," *Forbes*, April 23, 2018.

31. Costs are dropping toward around $340 to store 1 kWh. O. Schmidt et al., "The Future Cost of Electrical Energy Storage Based on Experience Rates," *Nature Energy* 2 (July 10, 2017). An installed Tesla Powerwall is about $563/kWh. www.tesla.com/powerwall. (That's $6,200 equipment cost plus "$800 to $2,000" for installation, for 13.5 kWh.) Prices will likely drop further in the coming years.
32. "Lazard's Levelized Cost of Storage Analysis, Version 3.0," November 2017, www.lazard.com/perspective/levelized-cost-of-storage-2017/, 12.
33. From about 4.5 cents/kWh to 8.2 cents. Ibid., 2.
34. *BP Statistical Review, 2017*, 46.
35. Brett Cuthbertson and Will Howard, "Backing Up the Planet—World Energy Storage," Office of the Chief Scientist, Australian Government, www.chiefscientist.gov.au/wp-content/uploads/Battery-storage-FINAL.pdf.
36. Geoffrey Smith, "Bill Gates Is Doubling His Billion-Dollar Bet on Renewables," *Fortune*, June 26, 2015.
37. "Lazard's Levelized Cost of Energy." See footnote 12.
38. Mark Z. Jacobson et al., "Low-Cost Solution to the Grid Reliability Problem with 100% Penetration of Intermittent Wind, Water, and Solar for All Purposes," *Proceedings of the National Academy of Sciences* 112, no. 49 (2015): 15060–15065.
39. Mark Z. Jacobson et al., "100% Clean and Renewable Wind, Water, and Sunlight All-Sector Energy Roadmaps for 139 Countries of the World," *Joule* 1, no. 1 (2017): 108–121.
40. Christopher T. Clack et al., "Evaluation of a Proposal for Reliable Low-Cost Grid Power with 100% Wind, Water, and Solar," *Proceedings of the National Academy of Sciences* 114, no. 26 (2017): 6722–6727; Eduardo Porter, "Traditional Sources of Energy Have Role in Renewable Future," *New York Times*, June 21, 2017, B1.
41. www.nytimes.com/interactive/2017/08/29/opinion/climate-change-carbon-budget.html.
42. Sanghyun Hong, Staffan Qvist, and Barry W. Brook, "Economic and Environmental Costs of Replacing Nuclear Fission with Solar

and Wind Energy in Sweden," *Energy Policy* 112 (January 2018): 56–66.

43. Swedish Television, "Misslyckat projekt med sol- och vindkraft i Simris," 2018, www.svt.se/nyheter/lokalt/skane/misslyckat-projekt-med-sol-och-vindkraft-i-simris. The Simris microgrid performance can be seen live here: https://les.eon.se. As of March 12, 2018, 83 percent of Simris electricity has been supplied by the national electricity grid and 17 percent from the renewable microgrid itself.
44. Johan Aspegren, head of communications, EON. See also Swedish Television, "Misslyckat projekt med sol- och vindkraft i Simris."
45. According to the Swedish state utility Vattenfall, the lowest lifecycle-emission sources in Sweden are nuclear power and hydroelectric. They make up more than 80 percent of the national grid production but none of the Simris microgrid production.
46. Paul Hawken et al., *Drawdown: The Most Comprehensive Plan Ever Proposed to Reverse Global Warming* (New York: Penguin, 2017), 220.
47. Ibid., 21.
48. Pew Research Center, Spring 2015 Global Attitudes Survey, question 84.
49. International Renewable Energy Agency, *Renewable Power Generation Costs in 2017* (Abu Dhabi: IRENA, 2018).

6. Methane Is Still Fossil

1. See www.uniongas.com/about-us/about-natural-gas/Chemical-Composition-of-Natural-Gas.
2. Keith Bradsher, "Even Spandex Is Hit by an Energy Squeeze," *New York Times*, December 13, 2017, B2.
3. Steven Lee Meyers, "In China's Coal Country, Shivering for Cleaner Air," *New York Times*, February 11, 2018, A5.
4. Nicholas Kawa, "Gas Leaks Can't Be Tamed," *Atlantic*, September 18, 2015.

5. Robert W. Howarth, "A Bridge to Nowhere: Methane Emissions and the Greenhouse Gas Footprint of Natural Gas," *Energy Science & Engineering* 2, no. 2 (2014): 47–60; Gayathri Vaidyanathan, "How Bad of a Greenhouse Gas Is Methane?," *Scientific American* (December 22, 2015).
6. P. J. Gerber et al., *Tackling Climate Change Through Livestock: A Global Assessment of Emissions and Mitigation Opportunities* (Rome: Food and Agriculture Organization of the United Nations, 2013); Matthew J. Vucko et al., "The Effects of Processing on the In Vitro Antimethanogenic Capacity and Concentration of Secondary Metabolites of *Asparagopsis taxiformis*," *Journal of Applied Phycology* 29, no. 3 (2017): 1577–1586.
7. M. Saunois et al., "The Growing Role of Methane in Anthropogenic Climate Change," *Environmental Research Letters* 11 (2016): 120207; Stefan Schwietzke et al., "Upward Revision of Global Fossil Fuel Methane Emissions Based on Isotope Database," *Nature* (October 6, 2016).
8. www.aljazeera.com/news/2017/10/blast-gas-station-rocks-ghana-capital-accra-171007211009348.html.

7. Safest Energy Ever

1. Associated Press, "Onagawa: Japanese Tsunami Town Where Nuclear Plant Is the Safest Place," *Guardian*, March 30, 2011.
2. "The results suggest that…it was not advisable to relocate any of the 162,700 actually relocated. This is because the inhabitants' gain in life expectancy, even in the most contaminated settlements…would have been insufficient to balance the fall in their life quality index caused by their notional payment of the costs of relocation." I. Waddington et al., "J-Value Assessment of Relocation Measures Following the Nuclear Power Plant Accidents at Chernobyl and Fukushima Daiichi," *Process Safety and Environmental Protection* 112 (2017): 35.
3. Koichi Tanigawa et al., "Loss of Life After Evacuation: Lessons Learned from the Fukushima Accident," *Lancet* 379 (March 10, 2012): 889–891.

4. A. Hasegawa et al., "Emergency Responses and Health Consequences After the Fukushima Accident: Evacuation and Relocation," *Clinical Oncology* 228 (2016): 237–244; Yuriko Suzuki et al., "Psychological Distress and the Perception of Radiation Risks: The Fukushima Health Management Survey," *Bulletin of the World Health Organization* 93 (2015): 598–605.
5. Hasegawa et al., "Emergency Responses and Health Consequences," 241.
6. Seth Baum, "Japan Should Restart More Nuclear Power Plants," *Bulletin of the Atomic Scientists* (October 20, 2015).
7. Lost nuclear generation in Japan and Germany after 2011 was about 400 TWh per year. In Japan about 21 percent of the lost nuclear power was replaced with coal and another 14 percent with oil. See www.enecho.meti.go.jp/en/category/brochures/pdf/japan_energy_2016.pdf. Death estimates are based on the formula in Staffan A. Qvist and Barry W. Brook, "Environmental and Health Impacts of a Policy to Phase Out Nuclear Power in Sweden," *Energy Policy* 84 (2015): 1–10. See also Mari Iwata, "Japan's Answer to Fukushima: Coal Power," *Wall Street Journal*, March 27, 2014; and Edson Severnini, "Impacts of Nuclear Plant Shutdown on Coal-Fired Power Generation and Infant Health in the Tennessee Valley in the 1980s," *Nature Energy* 2 (2017), article no. 17051.
8. David Ropeik, "The Dangers of Radiophobia," *Bulletin of the Atomic Scientists* 72, no. 5 (2016): 311–317.
9. The Chernobyl Forum (International Atomic Energy Agency et al.), *Chernobyl's Legacy: Health, Environmental, and Socio/Economic Impacts*, rev. ed. (Vienna: IAEA, 2006), 8.
10. Colin Barras, "The Chernobyl Exclusion Zone Is Arguably a Nature Preserve," BBC, April 22, 2016.
11. The number of justifiable evacuations is estimated at 9 percent to 22 percent of those actually evacuated. See Waddington et al., "J-Value Assessment of Relocation Measures."
12. Extrapolated from Gwyneth Cravens, *Power to Save the World: The Truth About Nuclear Energy* (New York: Vintage, 2007), 140–141.

13. Sammy Fretwell, "Santee Cooper Will Be Awash in Excess Power If SC Nuke Project Is Completed," *State*, July 19, 2017; Mark Nelson and Michael Light, "New South Carolina Nuclear Plant Would Cut Coal Use by 86%, New Analysis Finds," *Environmental Progress* (July 31, 2017).
14. "Nearly Completed Nuclear Plant Will Be Converted to Burn Coal," *New York Times*, January 2, 1984.
15. International Energy Agency, "Tracking Progress: Coal-Fired Power," 2017, www.iea.org/etp/tracking2017/coal-firedpower/.
16. Coal deaths for Europe are the average of estimates for lignite (the majority of coal in Europe), at thirty-three deaths per TWh and hard coal at twenty-five. The comprehensive European Union study *ExternE* is summarized in Anil Markandya and Paul Wilkinson, "Electricity Generation and Health," *Lancet* 370 (September 13, 2007): 979–990. The China estimate is from Eliasson Baldur and Yam Y. Lee, eds., *Integrated Assessment of Sustainable Energy Systems in China* (Dordrecht, Netherlands: Kluwer, 2003).
17. Duane W. Gang, "Five Years After Coal Ash Spill, Little Has Changed," *USA Today*, December 22, 2013.
18. www.greenpeace.org/archive-international/en/news/features/coal-ash-spills-expose-more-of/.
19. See note 16.
20. Extrapolated to 2017 from P. A. Kharecha and J. E. Hansen, "Prevented Mortality and Greenhouse Gas Emissions from Historical and Projected Nuclear Power," *Environmental Science & Technology* 47 (2013): 4889–4895.
21. Markandya and Wilkinson, "Electricity Generation and Health."
22. Ibid.
23. www.sfgate.com/news/article/Biggest-dam-failures-in-U-S-history-10928774.php.
24. David A. Graham, "How Did the Oroville Dam Crisis Get So Dire?," *Atlantic*, February 13, 2017; Mike James, "Tens of Thousands Evacuated Amid Failing Dam Crisis in Puerto Rico," *USA Today*, September 22, 2017.

25. Gloria Goodale, "Nuclear Radiation in Pop Culture: More Giant Lizards than Real Science," *Christian Science Monitor*, March 30, 2011.
26. Cravens, *Power to Save the World*, 72–73; Bruno Comby, *Environmentalists for Nuclear Energy* (1994; English translation, Paris: TNR Editions, 2001), 231–233.
27. M. Ghiassi-Nejad et al., "Very High Background Radiation Areas of Ramsar, Iran: Preliminary Biological Studies," *Health Physics* 82, no. 1 (2002): 87–93.
28. Cravens, *Power to Save the World*, 98.
29. Public Health England, "Ionising Radiation: Dose Comparisons," March 18, 2011, www.gov.uk/government/publications/ionising-radiation-dose-comparisons/ionising-radiation-dose-comparisons.
30. Cravens, *Power to Save the World*, 73.
31. Public Health England, "Ionising Radiation: Dose Comparisons."
32. A. D. Wrixon, "New ICRP Recommendations," *Journal of Radiological Protection* 28, no. 2 (2008): 161–168.
33. World Health Organization, *Health Risk Assessment from the Nuclear Accident After the 2011 Great East Japan Earthquake and Tsunami, Based on a Preliminary Dose Estimation* (Geneva: WHO, 2013).
34. L. E. Feinendegen, "Evidence for Beneficial Low Level Radiation Effects and Radiation Hormesis," *British Journal of Radiology* 78, no. 925 (2005): 3–7.
35. Angela R. McLean et al., "A Restatement of the Natural Science Evidence Base Concerning the Health Effects of Low-Level Ionizing Radiation," *Proceedings of the Royal Society B: Biological Sciences* (September 13, 2017); M. P. Little et al., "Risks Associated with Low Doses and Low Dose Rates of Ionizing Radiation: Why Linearity May Be (Almost) the Best We Can Do," *Radiology* 251 (2009): 6–12; M. Tubiana et al., "The Linear No-Threshold Relationship Is Inconsistent with Radiation Biologic and Experimental Data," *Radiology* 251 (2009): 13–22.

36. www.icrp.org/icrpaedia/effects.asp
37. www.grandcentralterminal.com/about. The figure of 750,000 for twenty minutes each is equivalent to 10,000 around the clock, which delivers 50,000 mSv per year in the aggregate at 5 mSv per year. With 1 percent cancer death risk for each 200 mSv (ICRP), the result is 2.5 fatalities yearly.
38. World Health Organization, *Health Risk Assessment*, 32, 59, 56.
39. Cravens, *Power to Save the World*, 228–229, 235; Nuclear Energy Institute, *Deterring Terrorism: Aircraft Crash Impact Analyses Demonstrate Nuclear Power Plant's Structural Strength* (Washington, DC: Nuclear Energy Institute, 2002).

8. Risks and Fears

1. David Ropeik, *How Risky Is It, Really? Why Our Fears Don't Always Match the Facts* (New York: McGraw-Hill, 2010); Steven Pinker, *The Better Angels of Our Nature: Why Violence Has Declined* (New York: Viking, 2011), 345–346; Scott L. Montgomery and Thomas Graham, Jr. *Seeing the Light: The Case for Nuclear Power in the 21st Century* (Cambridge: Cambridge University Press, 2017), 209–243.
2. Amos Tversky and Daniel Kahneman, "Availability: A Heuristic for Judging Frequency and Probability," *Cognitive Psychology* 5, no. 2 (1973): 207–232; Spencer R. Weart, *The Rise of Nuclear Fear* (Cambridge, MA: Harvard University Press, 2012).
3. G. Gigerenzer, "Dread Risk, September 11, and Fatal Traffic Accidents," *Psychological Science* 15, no. 4 (2004): 286–287.
4. Yoshitake Takebayashi et al., "Risk Perception and Anxiety Regarding Radiation After the 2011 Fukushima Nuclear Power Plant Accident: A Systematic Qualitative Review," *International Journal of Environmental Research and Public Health* 14, no. 11 (2017): 1306.
5. Robert Jay Lifton, "Beyond Psychic Numbing: A Call to Awareness," *American Journal of Orthopsychiatry* 52, no. 4 (1982): 619–629.

6. Similarly, more scientific information about climate change does little to change people's biases. See Dan M. Kahan et al., "The Polarizing Impact of Science Literacy and Numeracy on Perceived Climate Change Risks," *Nature Climate Change* 2 (October 2012): 732–735; Ezra Klein, "How Politics Makes Us Stupid," *Vox*, April 6, 2014.
7. Paul Slovic, "Perception of Risk," *Science* 236 (April 17, 1987): 280–285.
8. Paul Slovic, Baruch Fischhoff, and Sarah Lichtenstein, "Facts and Fears: Societal Perception of Risk," *Advances in Consumer Research* 8 (1981): 497–502.
9. Weart, *Rise of Nuclear Fear*, 188–189.
10. Andrew Newman, "The Persistence of the Radioactive Bogeyman," *Bulletin of the Atomic Scientists* 23 (October 2017).
11. David Ropeik, "The Rise of Nuclear Fear—How We Learned to Fear the Radiation," Scientific American blog, June 15, 2012.
12. See, for example, the photo of a demonstration against nuclear power illustrating an unrelated article about nuclear weapons: Max Fisher, "European Nuclear Weapons Program Would Be Legal, German Review Finds," *New York Times*, July 5, 2017.
13. There are many examples of the same phenomena; the *Titanic* did not end the operation of passenger ships.
14. "No New Record-Low for Road Deaths in Sweden," *Local*, January 9, 2017, www.thelocal.se/20170109/no-new-record-low-for-road-deaths-in-sweden; "Why Sweden Has So Few Road Deaths," *Economist*, February 26, 2014.
15. Reynold Bartel, Thomas Wellock, and Robert J. Budnitz, "WASH-1400, the Reactor Safety Study," Technical Report NUREG/KM-0010, U.S. Nuclear Regulatory Commission, August 2016.
16. "Scientists Criticize U.S. on Nuclear Safety Data," *New York Times*, November 18, 1977.
17. Alvin M. Weinberg, "A Nuclear Power Advocate Reflects on Chernobyl," *Bulletin of the Atomic Scientists* (August–September 1986): 57–60.

9. Handling Waste

1. Both coal and nuclear data from Gwyneth Cravens, *Power to Save the World: The Truth About Nuclear Energy* (New York: Alfred A. Knopf, 2007), 9.
2. See SKB.com.
3. It gained a critical, though not final, regulatory approval in 2018. Swedish Radiation Safety Authority, "Swedish Radiation Safety Authority Issues Pronouncement on Final Disposal," January 23, 2018, www.stralsakerhetsmyndigheten.se/en/press/news/2018/swedish-radiation-safety-authority-issues-pronouncement-on-final-disposal/.
4. http://posiva.fi/en; Elizabeth Gibney, "Why Finland Now Leads the World in Nuclear Waste Storage," *Nature* (December 2, 2015); "To the Next Ice Age and Beyond," *Economist* (April 15, 2017).
5. Rauli Partanen and Janne M. Korhonen, *Climate Gamble: Is Anti-nuclear Activism Endangering Our Future?*, 3rd ed. (n.p.: CreateSpace, 2017), 64–66.
6. Swedish Radio, "Slutförvar under Rönnskärsverken," 2017, http://sverigesradio.se/sida/artikel.aspx?programid=1650&artikel=6660810.
7. Boliden initially fought against the requirement to even store this material underground at all, saying that the "cost is not in proportion to the environmental benefits." Ny Teknik, "Boliden tar strid mot regeringen om kvicksilvret," 2003, www.nyteknik.se/digitalisering/boliden-tar-strid-mot-regeringen-om-kvicksilvret-6448341.
8. SKB, "Ny kostnadsberäkning för hanteringen av kärnavfallet," 2017, www.skb.se/nyheter/ny-kostnadsberakning-for-hanteringen-av-det-svenska-karnavfallet/.
9. www.andra.fr/international/.
10. http://nuclearsafety.gc.ca/eng/waste/high-level-waste/index.cfm; www.nwmo.ca/.
11. Extrapolated by ten years from Cravens, *Power to Save the World*, 269.

12. "Put Yucca Mountain to Work: The Nation Needs It" (editorial), *Washington Post*, July 15, 2017.
13. Ralph Vartabedian, "Nuclear Accident in New Mexico Ranks Among the Costliest in U.S. History," *Los Angeles Times*, August 22, 2016.
14. Cravens, *Power to Save the World*, 280–285.
15. www.nrc.gov/waste/spent-fuel-storage/faqs.html.
16. No health effects have resulted from the very rare minor accidents that have occurred. Kevin J. Connolly and Ronald B. Pope, "A Historical Review of the Safe Transport of Spent Nuclear Fuel," U.S. Department of Energy, August 31, 2016, FCRD-NFST-2016-000474, Rev. 1.

10. Preventing Proliferation

1. Isotopes are variants of chemical elements with different numbers of neutrons. Uranium in nature mostly has a large nucleus with 238 neutrons and protons (U^{238}), but less than 1 percent is the isotope U^{235}, with three fewer neutrons, which is much more likely to fission and allow a chain reaction.
2. William J. Broad, "From Warheads to Cheap Energy," *New York Times*, January 8, 2014, D1.
3. World Nuclear Association, "Military Warheads as a Source of Nuclear Fuel," updated February 2017, www.world-nuclear.org/information-library/nuclear-fuel-cycle/uranium-resources/military-warheads-as-a-source-of-nuclear-fuel.aspx.
4. Cravens, *Power to Save the World*, 148–152.
5. William C. Sailor et al., "A Nuclear Solution to Climate Change?," *Science* 288 (May 2000): 1178.
6. Nicholas L. Miller, "Why Nuclear Energy Programs Rarely Lead to Proliferation," *International Security* 42, no. 2 (2017): 40–77.
7. Byung-koo Kim, *Nuclear Silk Road: The Koreanization of Nuclear Power Technology* (n.p.: CreateSpace, 2011).
8. Robert H. Socolow and Alexander Glaser, "Balancing Risks: Nuclear Energy and Climate Change," *Dædalus* (Fall 2009): 31–44.

See also Steven E. Miller and Scott D. Sagan, "Nuclear Power Without Nuclear Proliferation" (special issue introduction), *Dædalus* (Fall 2009): 7–18.

9. www.nytimes.com/interactive/2015/03/31/world/middleeast/simple-guide-nuclear-talks-iran-us.html.
10. www.iaea.org/sites/default/files/the-iaea-leu-bank.pdf; Mariya Gordeyeva, "U.N. Nuclear Watchdog to Open Uranium Bank that May Have No Clients," Reuters, July 11, 2017.
11. Daniel B. Poneman, "The Case for American Nuclear Leadership," *Bulletin of the Atomic Scientists* 73, no. 1 (2017): 44–47.
12. www.world-nuclear.org/information-library/non-power-nuclear-applications/transport/nuclear-powered-ships.aspx.
13. Lower-scale or shorter armed conflicts (or both) between states, such as Russia-Georgia and Russia-Ukraine in recent years, do occur but are far less lethal than the sustained, high-level interstate wars of the past.

11. Keep What We've Got

1. Richard K. Lester and Robert Rosner, "The Growth of Nuclear Power: Drivers and Constraints," *Dædalus* (Fall 2009):19–30.
2. The nuclear production of 800 TWh/year is 57 percent of the nonfossil electricity. Reactor total is for February 2018.
3. www.nei.org/Knowledge-Center/Nuclear-Statistics/World-Statistics/World-Nuclear-Generation-and-Capacity; www.eia.gov/tools/faqs/faq.php?id=427&t=3.
4. Ann S. Bisconti, "Public Opinion on Nuclear Energy: What Influences It," *Bulletin of the Atomic Scientists* (April 27, 2016); Demoskop [polling company], "Rapport: Attityder till kärnkraften Ringhals," Vattenfall [utility], November 5, 2017, 14, https://corporate.vattenfall.se/globalassets/sverige/nyheter/attityder_till_ringhals_2010.pdf_16643410.pdf.
5. Bisconti, "Public Opinion on Nuclear Energy."
6. Nuclear Energy Institute, "Nuclear Power Plant Neighbors Accept Potential for New Reactor Near Them by Margin of 3 to 1,"

October 12, 2005, www.prnewswire.com/news-releases/nuclear-power-plant-neighbors-accept-potential-for-new-reactor-near-them-by-margin-of-3-to-1-55167507.html. However, the opposite may be true of *planned* plants, at least in China. See Yue Guo and Tao Ren, "When It Is Unfamiliar to Me: Local Acceptance of Planned Nuclear Power Plants in China in the Post-Fukushima Era," *Energy Policy* 100 (2017): 113–125.

7. Robert Surbrug, *Beyond Vietnam: The Politics of Protest in Massachusetts, 1974–1990* (Amherst: University of Massachusetts Press, 2009), 19–98.
8. Report to Vermont Department of Public Service on Vermont Yankee License Renewal, Chapter 12, 4, www.leg.state.vt.us/jfo/envy/7440%20Alternatives%20Report.pdf.
9. James Conca, "Who Told Vermont to Be Stupid?," *Forbes* (September 1, 2013); Scott DiSavino, "Massachusetts OK's Cape Wind / NSTAR Power Purchase Pact," Reuters, November 26, 2012.
10. Roger H. Bezdek and Robert M. Wendling, "A Half Century of US Government Energy Incentives: Value, Distribution, and Policy Implications," *International Journal of Global Energy Issues* 27, no. 1 (2007): 42–60, esp. 43. Of the nuclear power support, 96 percent was for research and development, whereas for fossil fuels only 8 percent was for R&D, with most of the rest being giveaways making operations cheaper. The limited support for renewables before 2003 was split between R&D and operations.
11. Jack (Anthony) Gierzynski, *The Vermont Legislative Research Service: Federal and Vermont State Subsidies for Renewable Energy* (Burlington: University of Vermont, 2016).
12. Bob Salsberg, "Massachusetts Taps Northern Pass for Hydropower Project," AP News, January 25, 2018; Michael Cousineau, "Northern Pass 'Shocked and Outraged' by Application Denial," *New Hampshire Union Leader*, February 1, 2018.
13. Vamsi Chadalavada, *Cold Weather Operations: December 24, 2017–January 8, 2018* (ISO New England [grid operator], January 16, 2018), www.iso-ne.com/static-assets/documents/2018/01/20180112_cold_weather_ops_npc.pdf.

14. Energy Information Administration data: www.eia.gov/state/print.php?sid=MA.
15. Mary C. Serreze, "Closure of Vermont Yankee Nuclear Plant Boosted Greenhouse Gas Emissions in New England," *Republican*, February 18, 2017, www.masslive.com/news/index.ssf/2017/02/report_closure_of_vermont_yank.html.
16. Electricity costs rose by $350 million and carbon emissions rose by 10 million tons. Lucas Davis and Catherine Hausman, "Market Impacts of a Nuclear Power Plant Closure," *American Economic Journal: Applied Economics* 8, no. 2 (2016): 120.
17. "Joint Proposal of Pacific Gas and Electric Company, Friends of the Earth, . . . to Retire Diablo Canyon Nuclear Power Plant," www.pge.com/includes/docs/pdfs/safety/dcpp/JointProposal.pdf.
18. Gwyneth Cravens, *Power to Save the World: The Truth About Nuclear Energy* (New York: Vintage, 2007), 247.
19. Rachel Becker, "New York City's Closest Nuclear Power Plant Will Close in Five Years," *Verge*, January 9, 2017.
20. Geoffrey Haratyk, "Early Nuclear Retirements in Deregulated U.S. Markets: Causes, Implications and Policy Options," *Energy Policy* 110 (2017): 150–166; Devashree Saha, "Nuclear Power and the U.S. Transition to a Low-Carbon Energy Future," Council of State Governments, July 7, 2017, knowledgecenter.csg.org/kc/content/nuclear-power-and-us-transition-low-carbon-energy-future.
21. World Nuclear Association, "Nuclear Power in Japan," May 24, 2017, www.world-nuclear.org.
22. These fossil imports drain the Japanese economy of at least $35 billion every year.
23. Geert De Clercq and Michel Rose, "France Postpones Target for Cutting Nuclear Share of Power Production," Reuters, November 7, 2017.
24. Meanwhile, in 2017, voters in Switzerland decided to ban new nuclear reactors but keep four existing reactors running (a fifth will close in 2019), while investing heavily in renewables. Michael

Shields and John Miller, "Swiss Voters Embrace Shift to Renewable Energy," Reuters, May 21, 2017.

25. Byung-koo Kim, *Nuclear Silk Road: The Koreanization of Nuclear Power Technology* (n.p.: CreateSpace, 2011). See especially Chapter 9 on standardization. World Nuclear Association, "Nuclear Power in South Korea," updated February 2017, www.world-nuclear.org.
26. Michael Shellenberger, "Greenpeace's Dirty War on Clean Energy, Part I: South Korean Version," *Environmental Progress* (July 25, 2017).
27. Sang-hun Choe, "In Reversal, South Korean President Will Support Construction of 2 Nuclear Plants," *New York Times*, October 21, 2017, A4.
28. Michael Shellenberger et al., "The High Cost of Fear: Understanding the Costs and Causes of South Korea's Proposed Nuclear Energy Phase-Out," *Environmental Progress* (August 2017).
29. Darrell Proctor, "Ringhals Delivers Record Output Despite Tough Economics," *Power* (November 2, 2017).
30. Sanghyun Hong, Staffan Qvist, and Barry W. Brook, "Economic and Environmental Costs of Replacing Nuclear Fission with Solar and Wind Energy in Sweden," *Energy Policy* 112 (January 2018): 56–66. See also F. Wagner and E. Rachlew, "Study on a Hypothetical Replacement of Nuclear Electricity by Wind Power in Sweden," *European Physical Journal Plus* 131 (2016): 173; and Staffan A. Qvist and Barry W. Brook, "Environmental and Health Impacts of a Policy to Phase Out Nuclear Power in Sweden," *Energy Policy* 84 (2015): 1–10.

12. Next-Generation Technology

1. The 1.2 GW VVER-1200.
2. Jessica Lovering, Loren King, and Ted Nordhaus, *How to Make Nuclear Innovative: Lessons from Other Advanced Industries*, Breakthrough Institute, March 2017, https://thebreakthrough.org/images/pdfs/How_to_Make_Nuclear_Innovative.pdf; Mark

Lynas, *Nuclear 2.0: Why a Green Future Needs Nuclear Power* (Cambridge: UIT Cambridge, 2014), 61–73; Richard K. Lester, "A Roadmap for U.S. Nuclear Energy Innovation," *Issues in Science and Technology* (Winter 2016): 45; Elisabeth Eaves, "Can North America's Advanced Nuclear Reactor Companies Help Save the Planet?," *Bulletin of the Atomic Scientists* 73, no. 1 (2017): 27–37.

3. Richard K. Lester, "A Roadmap for U.S. Nuclear Energy Innovation," *Issues in Science and Technology* (Winter 2016): 48.
4. Bill Gates, "Innovating to Zero!," TEDtalk, February 2010, www.ted.com/talks/bill_gates; Jason Pontin, "Q&A: Bill Gates," *Technology Review* (April 25, 2016).
5. Actually, the revised design essentially moves the fuel through a stationary wave. John Gilleland, Robert Petroski, and Kevan Weaver, "The Traveling Wave Reactor: Design and Development," *Engineering* 2, no. 1 (2016): 88–96.
6. Stephen Stapczynski, "Nuclear Experts Head to China to Test Experimental Reactors," *Bloomberg Technology* (September 21, 2017).
7. Richard Martin, *Superfuel: Thorium, the Green Energy Source for the Future* (New York: St. Martin's, 2012).
8. See lftrnow.com; thorconpower.com; and Robert Hargraves, *Thorium: Energy Cheaper Than Coal* (n.p.: CreateSpace, 2012).
9. J. Buongiorno et al., "The Offshore Floating Nuclear Plant Concept," *Nuclear Technology* 194, no. 1 (2016): 1–14.
10. Eric Ingersoll, personal communication, May 2018.
11. Dan Ariely, *Predictably Irrational: The Hidden Forces That Shape Our Decisions*, rev. ed. (New York: Harper, 2009), 1–22.
12. On bipartisan support for the Nuclear Energy Innovation and Modernization Act, see www.epw.senate.gov/public/index.cfm/neima. This bill passed in 2018.
13. iter.org/newline/-/2837.
14. Lev Grossman, "Fusion: Unlimited Energy. For Everyone. Forever," *Time*, November 2, 2015, 32–39. See also generalfusion.com.

15. Ashley E. Finan, "Strategies for Advanced Reactor Licensing," Nuclear Innovation Alliance, April 2016, www.nuclearinnovationalliance.org/advanced-reactor-licensing.
16. David Keith et al., "Stratospheric Solar Geoengineering Without Ozone Loss," *Proceedings of the National Academy of Sciences* 113, no. 52 (2016): 14910–14914; James Temple, "The Growing Case for Geoengineering," *Technology Review* 120, no. 3 (2017): 28–33.
17. Janos Pasztor, "Cooling-Off Period," *Technology Review* 120, no. 3 (2017): 10; James Temple, "China Builds One of the World's Largest Geoengineering Research Programs," *Technology Review* (August 2, 2017).
18. David Keith, *A Case for Climate Engineering* (Cambridge, MA: MIT Press, 2013).
19. www8.nationalacademies.org/onpinews/newsitem.aspx?RecordID=02102015.
20. James Temple, "Potential Carbon Capture Game Changer Nears Completion," *Technology Review* (August 30, 2017).

13. China, Russia, India

1. www.world-nuclear.org/information-library/current-and-future-generation/nuclear-power-in-the-world-today.aspx.
2. Edward Wong, "Coal Plants Threaten China's Climate Efforts," *New York Times*, February 8, 2017, A8.
3. Hiroko Tabuchi, "As Beijing Joins Climate Fight, Chinese Companies Build Coal Plants," *New York Times*, July 2, 2017, A10.
4. Junji Cao et al., "China-U.S. Cooperation to Advance Nuclear Power," *Science* 353, no. 6299 (2016): 547–548.
5. Jessica R. Lovering, Arthur Yip, and Ted Nordhaus, "Historical Construction Costs of Global Nuclear Power Reactors," *Energy Policy* 91 (April 2016): 371–382. For critical responses, see *Energy Policy* 102 (March 2017): 640–649.
6. International Energy Agency and OECD Nuclear Energy Agency, *Projected Costs of Generating Electricity, 2015 Edition* (Paris: IEA, 2015), 17, 41, 49, 83; Geoffrey Rothwell, "Defining Plant-Level

Costs. Presentation at OECD Workshop," Paris, January 20, 2016, 16; Geoffrey Rothwell, *Economics of Nuclear Power* (London: Routledge, 2015).

7. Also the CAP1000.
8. World Nuclear Association, "Nuclear Power in China," updated April 20, 2017, www.world-nuclear.org.
9. Stephen Chen, "Warships to Be Powered by Cold War Era Reactor," *South China Morning Post*, December 6, 2017.
10. Peter Fairley, "A Pyrrhic Victory for Nuclear Power," *IEEE Spectrum* (October 2017).
11. Matthew Cottee, "China's Nuclear Export Ambitions Run into Friction," *Financial Times*, August 2, 2017.
12. In ten locations, 27 GW total.
13. For this section, see World Nuclear Association, "Nuclear Power in Russia," updated July 27, 2017, www.world-nuclear.org/information-library/country-profiles/countries-o-s/russia-nuclear-power.aspx.
14. This role is not always welcomed in the United States. See Nick Gallucci and Michael Shellenberger, "Will the West Let Russia Dominate the Nuclear Market?," *Foreign Affairs* (August 3, 2017).
15. *Sentaku* Magazine, "Russia Unrivaled in Nuclear Power Plant Exports," *Japan Times*, July 27, 2017.
16. Information from World Nuclear Association supplemented by Daniel Westlén, personal communication.
17. World Nuclear Association, "Nuclear Power in Russia," "Transition to Fast Reactors" section.
18. Government of India, Ministry of Power, Central Electricity Authority, "Draft National Electricity Plan," vol. 1, December 2016, www.cea.nic.in/reports/committee/nep/nep_dec.pdf, 2.10.
19. Geeta Anand, "Until Recently a Coal Goliath, India Is Rapidly Turning Green," *New York Times*, June 3, 2017, A1.
20. Hans M. Kristensen and Robert S. Norris, "Indian Nuclear Forces, 2017." *Bulletin of the Atomic Scientists* 73, no. 4 (2017). India has air-, land-, and sea-based delivery systems.

21. To date, India has primarily constructed "heavy water" reactors, not the "light water" reactors mostly used around the world. (Light water is regular H_2O, while "heavy water" contains deuterium, which is a hydrogen isotope with an added neutron.) Heavy-water reactors can run on natural (not enriched) uranium, making them an ideal choice for nations cut off from access to the enriched uranium market.
22. "Way Forward Agreed for Jaitapur Reactors," *World Nuclear News* (March 12, 2018).

14. Pricing Carbon Pollution

1. Justin Gerdes, "How Much Do Health Impacts from Fossil Fuel Electricity Cost the U.S. Economy?," *Forbes*, April 8, 2013, www.forbes.com/sites/justingerdes/2013/04/08/how-much-do-health-impacts-from-fossil-fuel-electricity-cost-the-u-s-economy/#612d87edc679.
2. N. Gregory Mankiw, "A Carbon Fee That America Could Live With," *New York Times*, September 1, 2013, BU4.
3. U. Chicago Booth School, IGM Forum, "Carbon Tax," December 20, 2011, www.igmchicago.org/surveys/carbon-tax.
4. Eduardo Porter, "Counting the Cost of Fixing the Future," *New York Times*, September 11, 2013, B1.
5. William Nordhaus, *The Climate Casino: Risk, Economics, and Uncertainty for a Warming World* (New Haven, CT: Yale University Press, 2013), 177.
6. Ibid., 263.
7. Ibid., 225.
8. CDP North America, "Global Corporate Use of Carbon Pricing," September 2014, 10; Tamara DiCaprio, "The Microsoft Carbon Fee: Theory and Practice," Microsoft, December 2013; Georgina Gustin, "U.S. Rice Farmers Turn Sustainability into Carbon Credits, with Microsoft as First Buyer," *Inside Climate News* (June 26, 2017).
9. World Bank, Ecofys, and Vivid Economics, *State and Trends of Carbon Pricing, 2016* (Washington, DC: World Bank,

2016); Carbon Pricing Leadership Coalition, *Carbon Pricing Leadership Report, 2016–2017,* http://pubdocs.worldbank.org/en/183521492529539277/WBG-CPLC-2017-Leadership-Report-DIGITAL-Single-Pages.pdf.

10. James Temple, "Surge of Carbon Pricing Proposals Coming in the New Year," *Technology Review* (December 4, 2017).
11. www.worldbank.org/en/news/feature/2016/05/16/when-it-comes-to-emissions-sweden-has-its-cake-and-eats-it-too.
12. Eurostat, "Electricity and Heat Statistics," June 2017, http://ec.europa.eu/eurostat/statistics-explained/index.php/Electricity_and_heat_statistics.
13. Eduardo Porter, "British Columbia's Carbon Tax Yields Real-World Lessons," *New York Times,* March 2, 2016, B1.
14. www2.gov.bc.ca/gov/content/environment/climate-change/planning-and-action/carbon-tax.
15. https://ec.europa.eu/clima/policies/ets_en.
16. The share of free allowances is dropping toward 30 percent by 2020, although it remains above 80 percent in the aviation sector.
17. Alissa De Carbonnel, "Sweden Proposes Measures to Strengthen Carbon Prices," Reuters, October 17, 2016.
18. Dale Kasler, "California's Cap and Trade Program Is Costly, Controversial. But How Does It Work?," *Sacramento Bee,* July 19, 2017.
19. Chris Buckley, "China's Leader Pushes Ahead with Big Gamble on a Carbon Trading Market," *New York Times,* June 24, 2017, A4.
20. Keith Bradsher and Lisa Friedman, "China Plans Huge Market for Trading Pollution Credits," *New York Times,* December 20, 2017, B1.
21. Nordhaus, *Climate Casino,* 240.

15. Act Globally

1. Raymond Pierrehumbert, "How to Decarbonize? Look to Sweden," *Bulletin of the Atomic Scientists* 72, no. 2 (2016): 105–111.

2. Government of Ontario, "The End of Coal: An Ontario Primer on Modernizing Electricity Supply," November 2015, www.energy.gov.on.ca/en/files/2015/11/End-of-Coal-EN-web.pdf; Rauli Partanen and Janne M. Korhonen, *Climate Gamble: Is Anti-nuclear Activism Endangering Our Future?*, 3rd ed. (n.p.: CreateSpace, 2017), ix.
3. www.world-nuclear.org/information-library/country-profiles/countries-a-f/canada-nuclear-power.aspx.
4. www.energy.gov.on.ca/en/files/2015/11/End-of-Coal-EN-web.pdf.
5. Staffan A. Qvist and Barry W. Brook, "Potential for Worldwide Displacement of Fossil-Fuel Electricity by Nuclear Energy in Three Decades Based on Extrapolation of Regional Deployment Data," *PLoS ONE* 10, no. 5 (2015): e0124074; David Biello, "The World Really Could Go Nuclear," *Scientific American* (September 14, 2015); Pierrehumbert, "How to Decarbonize? Look to Sweden."
6. David Stanway, "Annual Nuclear Power Investment of $80 Billion Needed to Meet Climate Change Goals: IAEA," Reuters, April 27, 2017.
7. See Figure 57. South Korea's export reactors in the UAE cost about double the domestic ones.
8. John Mecklin, "Introduction: Nuclear Power and the Urgent Threat of Climate Change" (special issue), *Bulletin of the Atomic Scientists* 73, no. 1 (2017).
9. James Hansen et al., "Nuclear Power Paves the Only Viable Path Forward on Climate Change," *Guardian*, December 3, 2015; Dawn Stover, "Kerry Emanuel: A Climate Scientist for Nuclear Energy," *Bulletin of the Atomic Scientists* 73, no. 1 (2017): 7–12.
10. Sarah Booth Conroy, "Farewell Gestures," May 29, 1995, *Washington Post*, May 29, 1995.
11. Magdalena Andersson and Isabella Lövin, "Sweden: Decoupling GDP Growth from CO_2 Emissions Is Possible," World Bank blog Development in a Changing Climate, May 22, 2015, http://blogs.worldbank.org/climatechange/sweden-decoupling-gdp-growth-CO_2-emissions-possible.

12. David Roberts, "The Key to Tackling Climate Change: Electrify Everything," *Vox*, October 27, 2017.
13. www.ssab.com/company/sustainability/sustainable-operations/hybrit.
14. The fourth-generation "HTR-PM" reactor at Shiday Bay.
15. Jared Moore, "Thermal Hydrogen: An Emissions Free Hydrocarbon Economy," *International Journal of Hydrogen Energy* 30 (2017): 1–17.
16. Daisuke Miura and Tetsuo Tezuka, "A Comparative Study of Ammonia Energy Systems as a Future Energy Carrier, with Particular Reference to Vehicle Use in Japan," *Energy* 68 (April 2014): 428–436.
17. Robert Rosner and Alex Hearn, "What Role Could Nuclear Power Play in Limiting Climate Change?," *Bulletin of the Atomic Scientists* 73, no. 1 (2017): 2–6.
18. Framework Agreement between the Swedish Social Democratic Party, the Moderate Party, the Swedish Green Party, the Centre Party, and the Christian Democrats, June 10, 2016, www.government.se/49d8c1/contentassets/8239ed8e9517442580aac9bcb00197cc/ek-ok-eng.pdf.
19. Staffan A. Qvist and Barry W. Brook, "Environmental and Health Impacts of a Policy to Phase Out Nuclear Power in Sweden," *Energy Policy* 84 (2015): 1–10.
20. The figure is 805 TWh, 2016. www.nei.org/Knowledge-Center/Nuclear-Statistics/World-Statistics/Top-10-Nuclear-Generating-Countries.
21. Environmentalprogress.org; Breakthrough Institute, https://thebreakthrough.org; www.ecomodernism.org.
22. Meredith Angwin, *Campaigning for Clean Air: Strategies for Pronuclear Advocacy* (Wilder, VT: Carnot Communications, 2016); John Asafu-Adjaye et al., *An Ecomodernist Manifesto* (April 2015), www.ecomodernism.org/; Stewart Brand, *Whole Earth Discipline: An Ecopragmatist Manifesto* (New York: Viking, 2009); Joshua S. Goldstein and Steven Pinker, "Inconvenient Truths for the Environmental Movement," *Boston Globe*, November 23, 2015, A8.

23. Jessica Lovering et al., "Low-Carbon Portfolio Standards: Raising the Bar for Clean Energy. Breakthrough Institute and Environmental Progress," May 2016, thebreakthrough.org/index.php/issues/energy/low-carbon-portfolio-standards; Jared Moore, Kyle Borgert, and Jay Apt, "Could Low Carbon Capacity Standards Be More Cost Effective at Reducing CO_2 than Renewable Portfolio Standards?," *Energy Procedia* 63 (2014): 7459–7470.
24. Justin Gillis and Nadja Popovich, "The View from Trump Country, Where Renewable Energy Is Thriving," *New York Times*, June 8, 2017, A20.
25. Coral Davenport and Marjorie Connelly, "Half in G.O.P. Say They Back Climate Action," *New York Times*, January 31, 2015, A1.
26. Internationally, the politics of nuclear power are already very well developed. The IAEA and the NPT framework limit proliferation, along with technical measures administered by the Nuclear Suppliers Group. The World Association of Reactor Operators shares experience and information to improve safety. The World Nuclear Association coordinates the industry worldwide.
27. John Mueller, *Atomic Obsession: Nuclear Alarmism from Hiroshima to Al Qaeda* (Oxford: Oxford University Press, 2010).
28. http://environmentalprogress.org/global-overview.
29. Robert O. Keohane, "The Global Politics of Climate Change: Challenge for Political Science," *PS* 48, no. 1 (2015): 19–26; Robert O. Keohane and David G. Victor, "The Transnational Politics of Energy," *Dædalus* 142, no. 1 (2013): 97–109.
30. After large up-front licensing and construction costs, nuclear power plant operating expenses are lower than fossil-fuel plants, including methane, although higher than hydropower. See US Energy Information Administration, www.eia.gov/electricity/annual/html/epa_08_04.html.
31. Partanen and Korhonen, *Climate Gamble*, 78–89; Paul L. Joskow and John E. Parsons, "The Economic Future of Nuclear Power," *Dædalus* (Fall 2009): 45–47.
32. Ik Jeong and Lee Gye Seok, "ROK's Nuclear Policies and R&D Programs," presentation by Republic of Korea Ministry of

Science, ICT and Future Planning, at Nuclear Energy Agency International Workshop on the Nuclear Innovation Roadmap (NI2050), OECD, Paris, July 7–8, 2015.

33. World Nuclear Association, "Nuclear Power Economics and Project Structuring, 2017 Edition," www.world-nuclear.org, 4.
34. Ibid., 16. See also World Nuclear Association, "The Economics of Nuclear Power," updated April 2017, www.world-nuclear.org/information-library/economic-aspects/economics-of-nuclear-power.aspx.
35. World Nuclear Association, "Nuclear Power in South Korea," updated February 2017, www.world-nuclear.org. In the United States, Vogtle reactors 3–4, two AP1000s for $25 billion. www.utilitydive.com/news/vogtle-nuke-cost-could-top-25b-as-decision-time-looms/448555/. In the United Kingdom: "Hinkley Point: EDF Adds £1.5 Bn to Nuclear Plant Cost," *BBC News*, July 3, 2017.
36. Energy Technologies Institute, "The ETI Nuclear Cost Drivers Project: Summary Report," April 20, 2018. www.eti.co.uk/library/the-eti-nuclear-cost-drivers-project-summary-report.
37. However, innovation does also encounter resistance. See Calestous Juma, *Innovation and Its Enemies: Why People Resist New Technologies* (New York: Oxford University Press, 2016).
38. www.ecowatch.com/top-10-greenest-countries-in-the-world-1881962985.html; http://epi.yale.edu/sites/default/files/2016EPI_Full_Report_opt.pdf.
39. Alex Gray, "Why Sweden Beats Other Countries at Just About Everything," World Economic Forum website, January 30, 2017, www.weforum.org/agenda/2017/01/why-sweden-beats-most-other-countries-at-just-about-everything/.
40. Number one in the EU: http://ec.europa.eu/growth/industry/innovation/facts-figures/scoreboards_en. Number two worldwide: www.wipo.int/pressroom/en/articles/2016/article_0008.html.
41. www.weforum.org/agenda/2017/01/why-sweden-beats-most-other-countries-at-just-about-everything/; www.helpage.org/global-agewatch/population-ageing-data/global-rankings-map/.

Index

Index

Index

Index

Index

Index

Index

ALDEN COX

JOSHUA S. GOLDSTEIN is professor emeritus of International Relations at American University. He is the author of six books about war, peace, diplomacy, and economic history, and a bestselling college textbook, *International Relations*. See www.joshuagoldstein.com.

MONIKA JURASZEK

STAFFAN A. QVIST is a Swedish engineer, scientist, and consultant to clean energy projects around the world. Trained as a nuclear engineer, he works with both renewable and nuclear energy development projects. See www.staffanqvist.com.

PublicAffairs is a publishing house founded in 1997. It is a tribute to the standards, values, and flair of three persons who have served as mentors to countless reporters, writers, editors, and book people of all kinds, including me.

I. F. Stone, proprietor of *I. F. Stone's Weekly*, combined a commitment to the First Amendment with entrepreneurial zeal and reporting skill and became one of the great independent journalists in American history. At the age of eighty, Izzy published *The Trial of Socrates*, which was a national bestseller. He wrote the book after he taught himself ancient Greek.

Benjamin C. Bradlee was for nearly thirty years the charismatic editorial leader of *The Washington Post*. It was Ben who gave the *Post* the range and courage to pursue such historic issues as Watergate. He supported his reporters with a tenacity that made them fearless and it is no accident that so many became authors of influential, best-selling books.

Robert L. Bernstein, the chief executive of Random House for more than a quarter century, guided one of the nation's premier publishing houses. Bob was personally responsible for many books of political dissent and argument that challenged tyranny around the globe. He is also the founder and longtime chair of Human Rights Watch, one of the most respected human rights organizations in the world.

• • •

For fifty years, the banner of Public Affairs Press was carried by its owner Morris B. Schnapper, who published Gandhi, Nasser, Toynbee, Truman, and about 1,500 other authors. In 1983, Schnapper was described by *The Washington Post* as "a redoubtable gadfly." His legacy will endure in the books to come.

Peter Osnos, *Founder*